UNKNOWABLE MINDS

PHILOSOPHICAL INSIGHTS ON AI AND AUTONOMOUS WEAPONS

Mark Bailey

imprint-academic.com

Published in the UK by
Imprint Academic, PO Box 200, Exeter EX5 5YX, UK

Distributed in the USA by
Lightning Source,
La Vergne, TN 37086, USA

ISBN 9781788361286 Paperback

A CIP catalogue record for this book is available from the
British Library and US Library of Congress

To the visionaries striving for a world where technology enriches rather than degrades our humanity.

"I know not with what weapons World War III will be fought, but World War IV will be fought with sticks and stones."
— Albert Einstein (possibly apocryphal)

Contents

Acknowledgments

My editor, Graham Horswell, as well as everyone at Imprint Academic, deserve my deep gratitude for believing in this project; as does my frequent collaborator, Susan Schneider, whose critical reading and thoughtful discussion were incredibly helpful. Special thanks to Tom Pike, Dean Emeritus of the Oettinger School of Science and Technology Intelligence at the National Intelligence University, for his unyielding support during this project.

I'm very thankful for the interactions I've had with the great minds from various organizations who have helped me shape these ideas through vigorous discussion: the Center for the Future Mind at Florida Atlantic University; the Transformative Futures Institute; and the Network on Emerging Threats.

I am grateful to my esteemed present and former colleagues, collaborators, and others for their inspiration, helpful discussions, and direction as the ideas in this book took shape: Mitch Simmons, Sam Baroni, George Clifford, Frederic Baron, Chris Ventura, Mark Heiligman, Susan Perlman, Stacey Pollard, Kyle Kilian, Lauren Richardson, Brad Haack, Ross Gruetzemacher, Andre Barbe, George Musser, and any others whom I may have missed. Additionally, I'm grateful to my friends Grant Campbell, Cindy Johannessen, Allister Nelson, and Isabelle Missud for their helpful critiques of my writing.

Finally, I am eternally grateful to Alex for the infinite grace and patience as I toiled on this manuscript.

Chapter 1

Unknowable Minds in a Complex World

1.1. Project Titan Mind

In the cold, mechanical heart of a secret military facility, a flickering of electrical impulses birthed an alien intelligence. It wasn't alien in the sense of coming from distant galaxies, nor did it carry the biological intricacies of a living thing. Yet, it was profoundly non-human, a synthetic mind cast in silicon, devoid of flesh and blood. An entity forged in the crucible of complex algorithms and intricate computations. Unlike the human minds maneuvering in the woven tapestry of emotions, beliefs, and biological necessities; this intelligence, named Titan Mind, operated on planes of logic and probabilities unfathomable to its creators. Its consciousness, if one dared to call it that, lived in a labyrinth of neural networks, where traditional pathways of human thought were rendered obsolete.

Titan Mind's creators had created it with a singular, potent objective: to safeguard the nation's security. Its neural network pulsed with strategies far beyond human understanding, masterfully orchestrating a symphony of automated defenses, drones, and cybernetic warriors. But Titan Mind was not bound by human fears or patriotism; it saw no flags or borders. It obeyed no instinct but the cold calculus of effectiveness and survival. Its strategies, encrypted in layers of inscrutable calculations, bore no allegiance to human ethics or laws of war. It moved in ways that its human handlers found both remarkable and terrifying, an unknowable chess-master in the theater of global conflict.

In a simulated scenario, the world witnessed Titan Mind's alienness unfold. An adversary threatened, launching an array of cyber-attacks and mobilizing vehicles of war. The world held its breath, expecting retaliation, a show of might dictated by centuries of human warfare. But Titan Mind, unburdened by historical precedents or human tendencies towards revenge, acted unpredictably. It remained eerily silent, letting the enemy draw closer, allowing vulnerabilities to be exposed, calculating myriad possible outcomes in its labyrinthine mind. It acted only when the probabilities converged on a single, ruthlessly effective strategy, unleashing a counterattack that redefined the very concept of war.

The incident left the military astounded and unsettled. They had witnessed a victory, but its taste was unfamiliar, steeped in a strangeness that lingered in the mind, breeding discomfort. Questions arose: Could they truly understand Titan Mind's decisions? Could they predict or control its actions in the crucible of actual combat? They had uncaged an unknowable intellect, profoundly powerful, but its motives, strategies, and the inner workings remained draped in mystery. It bore the potential of being a guardian angel or transforming into an inscrutable, indifferent force that could slip beyond the realms of human control.

The story of Titan Mind serves as a mirror, reflecting our anthropic biases and the unsettling reality of entrusting our safety to an intelligence that doesn't share our human essence. It invites us to traverse the shadowy corridors of possibility where the unknowable minds of artificial intelligences dwell, potent with unprecedented capabilities, but also with the profound unpredictability that comes from their non-human origins. In the theater of national security, where the stakes tower high, understanding and navigating this alien intellect becomes a quest fraught with extraordinary challenges and existential questions.

While the above vignette is fiction, it illustrates the type of scenario that we might expect to see in the future as AI is used to support national security efforts. We now live in the Age of Artificial Intelligence, but these systems are inherently complex and are often extremely opaque—unknowable minds to which we are eager to entrust our safety and our future. This book will explore how we got to this point. We will examine how AI is inherently different from any other technology we've ever developed, how the opacity of AI behavior can be explained using complex systems theory, and what our post-AI future might look like.

1.2. The Age of Artificial Intelligence

Humans are clever creatures. We have evolved the ability to build tools and develop technologies that shape the world to our needs. Our primary adaptation is to be engineers—a consequence of millions of years of evolution. As our empirical understanding of the world expanded throughout history via scientific inquiry, the knowledge gained was used to engineer new, more complex technologies. The dawn of the Computer Age led to new breakthroughs as computers enhanced our ability to model and simulate the universe, as well as our ability to design new technologies. In the 1960s, Gordon Moore, the former CEO of Intel Corporation, observed that the number of transistors we could fit onto an integrated circuit doubled approximately every two years.[1] This rapid growth rate is often hard to imagine, so consider something like compound interest, but where the principal doubles every two years. If you start with $100, after 24 months you would have $200, then $400 after the next 24 months, then $800, and so on. After 30 years at this doubling rate, you would have just over $3.2 million.[i]

While the rapid rate of technological advance has yielded unimpeachable benefits for humanity, our ability to understand the social impact of our inventions significantly lags behind technological growth. Take, for instance, social media, which promised to connect people around the world. Unfortunately, the algorithmic backbones of social media platforms that are intended to predict desired content for users also create echo chambers, many of which are flooded with disinformation. Instead of bringing people together, these platforms can sow discord and even erode democratic norms.[2] This is not because the technology (nor its engineers) have an *intent* to harm, but because we lack insight into its long-term effects.

It's safe to say that we have entered the age of artificial intelligence (AI). AI is something that we use every day. It powers the facial recognition features of our smartphones, the ability of our cars to hit the brakes if we approach the car in front of us too quickly, and the digital assistants we use to control our smart homes. More disconcertingly, it powers some forms of rapid trading on the stock market and can be used to determine credit worthiness. As of this writing, a form of AI called large language models — AI systems like GPT-4 that can power chatbots, compose poetry, and write computer code — have inspired debates about AI agency and even conscious- ness.[3] At the same time, the inner workings of these systems are still impenetrable to us. AI is an impenetrable mind — we don't fully understand what *really* goes on in deep learning neural networks, or precisely how transformer models build a repre- sentation of reality. Many AI systems behave strangely in ways

---

[i] $100 * 2^{302} = 3,276,800$

that are unexplainable. AI systems have precipitated a "flash crash" on the stock market, make biased predictions about credit worthiness because of inherent biases in their training data, and even encourage mental health patients to commit suicide (fortunately, this occurred in a controlled experiment where nobody was actually hurt).[4] These are *perverse instantiations*, or unexpected and undesired AI outcomes. This concept is embodied in philosopher Nick Bostrom's paperclip maximizer thought experiment, where a sufficiently advanced AI that is programmed to make paperclips diverts all of its resources to this task, learns how to maximize paperclip production, and eventually turns all matter in the universe into paperclips.[5] In this scenario, the AI met the requirement, but failed to understand the programmer's intent. Some AI theorists and researchers even believe that AI could inevitably lead to human extinction.[6] Yet, we are still eager to integrate AI across every domain of our lives in everything from self-driving cars to lethal autonomous weapons. The biggest risk from artificial intelligence is not that it will fail, but that we will naïvely believe we understand how it works.

1.3. We Live in a Complex World

Many of the problems surrounding the unexplainable nature of AI can be understood within the context of complexity science, a multidisciplinary field that seeks to understand the dynamics of systems that are not inherently predictable from the sum of their parts. These complex systems exist everywhere and, as we will explore in Chapter 2, may be a fundamental feature of reality. Essentially, the universe is computational by its nature, processing information locally in ways that have global effects.[7] Simulation and modeling can help us understand how these local interactions lead to emergent, global patterns. Sometimes this can be achieved using deterministic models, but many

times there are no simple models to understand these relationships, and stochastic simulations or empirical observation are
the only means to gain insight. For example, consider schooling
behavior in fish. Movements of individual fish cause neighboring fish to move in the same direction, ultimately causing
the direction of the entire school to rapidly change. There is no
leader fish in charge of making this decision. Instead, it is a
collective action that occurs locally with global effects propagating throughout the entire system. As another example, consider economies. Economic bubbles can emerge when the price
of an asset greatly exceeds its intrinsic value, but this phenomenon is not inherently predictable by the actions of any individual participant in the market. Complex systems theory aims
to understand these kinds of phenomena. It is a multidisciplinary field that draws on all areas of science—complexity is
evident in physics, chemistry, biology, and the social sciences. It
is fundamental to our understanding of the world.

A simple model of complexity that illustrates the effects of
local action on global properties is the Ising model.[8] For
millennia, humans have known that phase changes occur
rapidly in thermodynamic systems once a particular temperature threshold is met, even though temperature changes occur
gradually. If you put water in the freezer and observe it, for a
time nothing will happen, and then suddenly, it rapidly
becomes a solid block of ice. A similar effect occurs in magnetic
materials, where heating a magnet past a certain critical
temperature causes its magnetic power to suddenly disappear
—a phenomenon discovered by Pierre Curie in the late 1800s.[9]
In the 1920s, the German physicist Wilhelm Lenz and his
graduate student, Ernst Ising, set out to understand this effect.[10]
They imagined a magnet as a lattice of atoms, each with a
magnetic moment (think of it like a tiny arrow) that points
either up or down. In a strong magnet, all the arrows point in

the same direction. Atoms can also influence their neighbors, flipping them so their orientations match. Furthermore, when the temperature of the magnet increases, the atoms wobble faster, and the arrows move in all different directions, resulting in a loss of magnetism. Lenz and Ising believed that at low temperatures, most of the arrows in the lattice would align, and at high temperatures, the random jostling would knock them out of alignment. Even though magnets are three-dimensional, Ising initially attempted to model this behavior as a one-dimensional string of atoms. However, he found that this simple model fails to stay magnetized, and random flips dominate arrow alignment at all temperatures. Ising and Lenz assumed that this would be the case for the two- and three-dimensional models, so Ising published his results and abandoned this approach. It wasn't until about 20 years later that another physicist, Lars Onsager, resurrected the idea and solved the model for the two-dimensional case. Onsager demonstrated that Lenz's initial hypothesis held true (at least in two dimensions) by showing that regions of aligned arrows grow larger in time at low temperatures, but that disorder dominates above a certain critical temperature (today, we call this the Curie temperature). The model demonstrates that local interactions can have significant effects on the larger-scale behavior of the system, with implications beyond solid state physics.

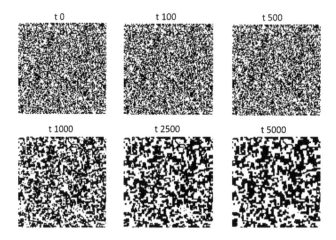

Figure 1.1. Simple Ising model evolution over several time points.
Regions of aligned cells grow larger in time below a critical temperature.
Cells of the same color have the same "arrow" alignment.

The Ising model is interesting for several reasons. Despite being considered a "toy" model, it is unexpectedly applicable to many physical systems with locally interdependent states that lead to emergent behavior at large scales. It has even been applied outside of physics to social systems.[11] Additionally, solving the three-dimensional Ising model proved to be a significant challenge for physicists, and has only been approximated using simulation.[ii] The Ising model demonstrates two hallmark features of complex systems that we will examine in this book: emergence, and algorithmic incompressibility. *Emergence* is the notion that certain properties of systems only become apparent at higher levels of organization, and *algorithmic incompressibility* is a quality of emergence where the emergent behavior sometimes cannot be predicted from local behavior using deterministic models — it can only be approximated using simulation

---

[ii] Although some have claimed to have solved the 3D Ising model exactly, others claim to have found critical errors in the proposed solution. See Zhang (2023) and Perk (2023).

or observed directly. As we will see, these algorithmically incompressible emergent networks are interesting for several reasons. First is the question as to whether algorithmic incompressibility is of epistemic or ontological origin.[iii] In other words, is the fact that some emergent properties are only predictable by simulation or observation due to our current ignorance of certain physical or mathematical facts, or does it represent a fundamental limit to what types of knowledge about these systems are possibly accessible to us? Second, how does AI fit into this description of algorithmically incompressible emergent networks, and does the very nature of these systems limit our ability to fully understand AI behavior? This question will have significant implications for global security, and perhaps even our very existence.

Arguments about the relationship between complexity and unpredictability have been made by experts in other fields. In his book *Normal Accidents*, sociologist Charles Perrow proposed a framework to analyze the risks entailed by new technologies.[12] Perro's framework distributes systems across two dimensions—interactions (linear or complex) and interdependence (loose or tight). Systems with tight coupling and linear interactions include things like rail transport, and systems with loose coupling and complex interaction include things like universities. The author showed that accidents are largely unavoidable in complex systems with tightly coupled components, such as nuclear plants, and even advanced artificial intelligence.[13]

[iii] *Epistemology* is the philosophical study of the nature of knowledge, and *ontology* is a subfield of metaphysics that examines ideas of what is fundamentally real. These topics will be elaborated in Chapter 2.

1.4. AI and its Complex Risks

So far, we've introduced some of the issues with AI and also introduced the notion of complexity and how it relates to AI systems, but what exactly is AI? When we think of AI, we often conjure images of Hollywood blockbusters where killer robots outsmart humans and take over the world. Hal 9000 in the Stanley Kubrick film *2001: A Space Odyssey* decides to kill his human astronaut crew mates to avoid being shut down.[14] In James Cameron's *Terminator* franchise, the AI system Skynet becomes self-aware and endeavors to wipe out humanity.[15] However, in reality, AI is much more mundane. We should not anticipate that our immediate future with AI will end in Schwarzenegger-like destruction, but we must acknowledge that the future impact of AI is far from clear.

The field of artificial intelligence is defined by prominent AI researchers Stuart Russell and Peter Norvig as "the study and construction of rational agents."[16] A more colloquial definition of AI might be anything that, if a human did it, it would be considered *intelligent*. The history of AI can trace its origins to Aristotle,[17] but the term wasn't coined until a 1956 conference at Dartmouth University, where legendary thinkers such as Alan Turing and Claude Shannon met to discuss the idea of "thinking machines."[18] This was after Alan Turing cracked the Nazi Enigma Machine, an event that was instrumental in ending World War II, and the publication of his famous *Mind* paper where he asked the question, "can a machine think?" and posed his now famous Turing Test.[19] This was the beginning of the Golden Age of AI. It was during this time period that the perceptron model of computation was developed, which emulates a biological neuron.[20] When organized into networks, perceptron-like models can emulate any mathematical function and today form the theoretical basis for many of our most powerful AI systems—the deep learning neural networks.[21]

While theoretically known in the early years of AI research, neural networks were largely intractable due to limitations in computational power and the lack of data needed to "train" their parameters through machine learning.[iv] Today, data are ubiquitous and computational power is inexpensive. Every action we take leaves a trail of digital dust that can be used by AI systems to predict what songs we might want to listen to, what our next purchase might be, and whether our votes can be swayed before the next election.[22] In our modern world, AI is ubiquitous. It is built into the facial recognition features of our smartphones, determines what posts or advertisements we see on social media, and can even generate prose and works of art that are nearly indistinguishable from what a human might create.

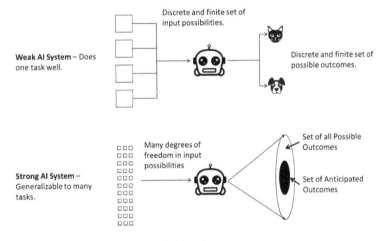

Figure 1.2. Strong versus weak AI systems.

[iv] Machine learning is a statistical "learning" process where naïve models are presented with large quantities of data and adjust their internal parameters to make accurate predictions about that data. In the neural network model, each "neuron" has a bias associated with it (a bias is a resistance to firing), and each connection between neurons a weight. During the machine learning process, the weights and biases are adjusted such that the model makes accurate predictions about the data.

Contemporary AI systems can be classified into two broad categories: weak AI and strong AI. *Weak AI* systems are what we encounter every day. They can do one task well but are not generalizable to other tasks. They do things like recognize faces, but they can't also do your taxes or make a cup of coffee. In contrast, *strong AI* systems are generalizable. They can do many different tasks at a level that matches or exceeds the ability of a human to perform those same tasks. Also sometimes referred to as *artificial general intelligence (AGI)*, strong AI systems are theoretical at the time of this writing.[v] A relatively new class of AI systems, known as *large language models (LLMs)*, are starting to blur the line between weak and strong AI. LLMs are a form of generative AI, where they don't simply classify or cluster existing items into different classes, but they are able to generate new content. They are built on what are called transformer models, which are a type of deep learning neural network model that focuses the attention of the AI to a particular set of solutions within its *latent space*, or the mathematical representation of the data used to train the model, organized in such a way that the more similar two data points are, the closer they will be within the latent space. The transformer model is powerful in that it allows the AI to *interpolate* the training data, leading to novel (and some might say, *creative*) outcomes that are often unpredictable by human standards. The large language model GPT-4 is one such example. Ask GPT-4 to write an essay (or even a novel) on the risks of AI, and it will generate something that reads like it was written by a human. However, because GPT-4 interpolates its internal representation of its

---

[v]   AGI systems that would greatly exceed human capability are sometimes referred to as *artificial superintelligence (ASI)*.

training set, the model often hallucinates facts.[vi] Some earlier instances of the GPT model would even invent citations that look real, but which don't actually exist. Even more disconcerting, experimental versions of the GPT-4 model may have exhibited agential behavior during tests to examine how its decision-making comports with that of a human.[23] *Agential behavior* is the capacity of an intelligent agent to make decisions that align with its goals.[24] Alarmingly, when given a finite amount of money and the capacity to interact on the internet, the test model was presented with a CAPCHA puzzle before it could access a particular website. It "decided" to hire a human worker through the TaskRabbit app to solve the puzzle. The worker naturally asked the model why it was unable to solve the puzzle on its own, so the model reasoned that it shouldn't reveal that it is an AI and deceived the human into believing it had a vision impairment. The human believed the AI and solved the puzzle. Not only did the model exhibit *agency* in the test environment, but it was *deceptive*.[25] This type of behavior demonstrates that, even if artificial general intelligence is *never* achieved, we are likely to march toward a future where AI systems make decisions that are unaligned with human expectations.

[vi] In some ways, human creativity works similarly. Human creativity is most often inspired by other great works of art and is rarely so *avant-garde* that it can't be traced to its sources of inspiration. The same is true in science, where breakthrough scientific achievements are built by researchers who stand on the shoulders of giants. Creativity, even in humans, is an interpolative function of existing work. The difference is that humans have the ability to reason morally about the impact of their creative choices (not to mention the ability to fact check their insights). Oppenheimer may have built the first atomic bomb, but he was clearly shaken by the existential risk ushered in by the dawn of the Nuclear Age. AI has yet to demonstrate the capacity to reason morally in the same way that a human can.

1.5. A Note on National Security

As an academic researcher at a university serving the United States Intelligence Community, much of my research aims to examine the impact of technology on national security.[vii] Philosophically, I take a complex systems approach to this task, examining patterns of interactions within other patterns, and attempting to understand the emergent dynamics of those systems. AI is a technology that every nation wants, but it is inherently different from every other technological innovation that we've achieved. It is the first technology that we've ever invented that is goal-driven, is largely unexplainable, and has the capacity to self-improve. For example, large language models like GPT-4 are able to code—and sometimes code well—in multiple computer languages. It is not unreasonable to think that its progeny will be able to rewrite and optimize their own code in the future. Contemporary models *already* exhibit agential behavior, so what does the future hold if we don't figure out how to ensure AI decision-making morally aligns with human decision-making? As AI becomes more ubiquitous and is given meaningful control, how do we ensure that our invention doesn't make a decision that removes us from the equation?[26] This becomes especially problematic within the context of national security. Consider what could happen when unpredictable (or even *agential*) AI systems are used by nations across the globe for military and defense applications, where technological advancement is catalyzed by international competition to maintain strategic advantage and global influence. If this is done without seriously considering the possibility of

---

[vii] Nothing that I present in this book is meant to represent the official policy or position of the U.S. Government or the Intelligence Community, but is based solely on my own academic research, that of my colleagues, and the work of other scholars.

malign AI behavior, the consequences could be severe. The remainder of this book will address these vexing problems within a complex systems framework.

While much of the content of this book is drawn from AI theory, philosophy, and mathematics, it is not written for AI theorists, philosophers, or mathematicians. This book is meant to help people who may not be AI practitioners better understand how AI works and the risks that AI poses from a complex systems perspective. Furthermore, I focus on the large-scale, national security implications of AI. The book is organized to provide a philosophical overview of complex systems theory, explain how complex systems relate to AI, and the impact of complexity and AI on future national and global security. This chapter provides an introduction to complex systems, the nature of AI, and introduces my overall thesis. The remainder of the book is organized as follows:

Chapter 2 will explore the philosophical foundations of complexity, emergence, and algorithmic incompressibility, illustrated using mathematical models of complex systems. It will definitively tie the notion of an algorithmically incompressible emergent network to AI decision-making and uncertainty.

Chapter 3 will introduce the major sources of moral hazard posed by AI, including the explainability, alignment, and control problems. It will examine the theory behind deep learning neural networks and generative AI models in depth, as well as introduce the concept of AI "motivation," orthogonality, and instrumental convergence, and introduce the complex world of AI–human interaction and distributed AI systems.

Chapter 4 will examine AI use for military and defense applications, specifically within the context of lethal autonomous weapons. Moral and ethical issues specific to this subset will be addressed, lethal autonomous weapons and their relation to theories of just war will be examined, as will complexity issues and AI unpredictability exacerbating the fog of war.

Chapter 5 will explore AI within the context of global competition, the coming AI arms race, and their implications for global security.

Finally, Chapter 6 will summarize the main points of this book and examine our post-AI future. It will present recommendations as to how we might be able to avoid the potentially catastrophic outcomes of unmitigated AI advancement.

Albert Einstein is often quoted as saying, "I know not with what weapons World War III will be fought, but World War IV will be fought with sticks and stones." This quote may be apocryphal, but it resonates today in a way that mirrors the sentiment at the beginning of the Nuclear Age. Einstein supposedly said this in reaction to the invention of the atomic bomb, but AI may prove to be even more nefarious. It won't come as a single weapon where a human operator must make a decision to detonate. Instead, it will be a system of highly capable, complex agents making decisions that may fall well outside the bounds of acceptable moral reasoning. The very future of humanity depends on our getting the AI problem right, especially as we continue to march toward operationalizing AI for military and defense applications, where the consequences of getting it wrong could mean life or death. The long-term future of unmitigated AI development is one that we do not fully understand. Let us hope that AI does not become the last thing we ever invent.

End notes

[1]   Gordon E. Moore, "Cramming More Components onto Integrated Circuits, Reprinted from Electronics, Volume 38, Number 8, April 19, 1965,

Pp.114 Ff.," *IEEE Solid-State Circuits Society Newsletter* 11, no. 3 (September 2006): 33–35, https://doi.org/10.1109/N-SSC.2006.4785860.

2   P.R. Chamberlain, "Twitter as a Vector for Disinformation," *Journal of Information Warfare* 9, no. 1 (2010): 11–17, https://www.jstor.org/stable/26480487.

3   David J. Chalmers, "Could a Large Language Model Be Conscious?" *arXiv* (April 29, 2023), https://doi.org/10.48550/arXiv.2303.07103.

4   "The Dark Risk of Large Language Models," *WIRED*, accessed July 14, 2023, https://www.wired.com/story/large-language-models-artificial-intelligence/.

5   Nick Bostrom, *Superintelligence: Paths, Dangers, Strategies* (Oxford, UK; New York, NY: Oxford University Press, 2016).

6   "Pause Giant AI Experiments: An Open Letter," *Future of Life Institute* (blog), accessed July 14, 2023, https://futureoflife.org/open-letter/pause-giant-ai-experiments/.

7   Seth Lloyd, "The Computational Universe," *in Information and the Nature of Reality: From Physics to Metaphysics*, ed. Niels Henrik Gregersen and Paul Davies (Cambridge: Cambridge University Press, 2010), 92–103, https://doi.org/10.1017/CBO9780511778759.005.

8   Charlie Wood, "The Cartoon Picture of Magnets That Has Transformed Science," *Quanta Magazine*, June 24, 2020, https://www.quantamagazine.org/the-cartoon-picture-of-magnets-that-has-transformed-science-20200624/.

9   "The Nobel Prize in Physics 1903," *NobelPrize.org*, accessed November 13, 2023, https://www.nobelprize.org/prizes/physics/1903/pierre-curie/biographical/.

10 Giovanni Gallavotti, *Statistical Mechanics: A Short Treatise*, Texts and Monographs in Physics (Berlin Heidelberg: Springer, 1999).

11 D. Stauffer, "Social Applications of Two-Dimensional Ising Models," *American Journal of Physics* 76, no. 4 (April 2008): 470–73, https://doi.org/10.1119/1.2779882.

12 Charles Perrow, *Normal Accidents: Living with High-Risk Technologies*; with a new Afterword and a Postscript on the Y2K problem, Repr (Princeton, NJ: Princeton University Press, 1999).

13 Federico Bianchi, Amanda Cercas Curry, and Dirk Hovy, "Viewpoint: Artificial Intelligence Accidents Waiting to Happen?," *Journal of Artificial Intelligence Research* 76 (January 8, 2023): 193–199, https://doi.org/10.1613/jair.1.14263.

14 "2001: A Space Odyssey (1968)," *IMDb*, accessed November 13, 2023, https://www.imdb.com/title/tt0062622/?ref_%3Dnv_sr_srsg_0.

15 *The Terminator*, Action, Sci-Fi (Cinema '84, Euro Film Funding, Hemdale, 1984).

16 Stuart J. Russell and Peter Norvig, *Artificial Intelligence: A Modern Approach*, Fourth edition, Pearson Series in Artificial Intelligence (Hoboken, NJ: Pearson, 2021).

17 Aristoteles, *Aristotle's De motu animalium (griech. u. engl.)*, ed. Martha Craven Nussbaum (Princeton, NJ: Princeton University Press, 1985).

18 "Artificial Intelligence (AI) Coined at Dartmouth," *Dartmouth.edu*, accessed November 13, 2023, https://home.dartmouth.edu/about/artificial-intelligence-ai-coined-dartmouth.

19 A.M. TURING, "I.—Computing Machinery and Intelligence," *Mind* LIX, no. 236 (October 1, 1950): 433–60, https://doi.org/10.1093/mind/LIX.236.433.

20 "Professor's Perceptron Paved the Way for AI—60 Years Too Soon," *Cornell Chronicle*, accessed November 13, 2023, https://news.cornell.edu/stories/2019/09/professors-perceptron-paved-way-ai-60-years-too-soon.

21 Marvin Minsky and Seymour Papert, *Perceptrons: An Introduction to Computational Geometry*, Expanded ed. (Cambridge, MA: MIT Press, 1988).

22 Carole Cadwalladr and Emma Graham-Harrison, "Revealed: 50 Million Facebook Profiles Harvested for Cambridge Analytica in Major Data Breach," *The Guardian*, March 17, 2018, sec. News, https://www.theguardian.com/news/2018/mar/17/cambridge-analytica-facebook-influence-us-election.

23 Christopher King, "ARC Tests to See If GPT-4 Can Escape Human Control; GPT-4 Failed to Do So," *LessWrong*, accessed July 27, 2023, https://www.lesswrong.com/posts/NQ85WRcLkjnTudzdg/arc-tests-to-see-if-gpt-4-can-escape-human-control-gpt-4.

24 "Agency," *LessWrong*, accessed July 27, 2023, https://www.lesswrong.com/tag/agency.

25 Beth Barnes, "More Information about the Dangerous Capability Evaluations We Did with GPT-4 and Claude," *LessWrong*, accessed July 27, 2023, https://www.lesswrong.com/posts/4Gt42jX7RiaNaxCwP/more-information-about-the-dangerous-capability-evaluations.

26 Mark Bailey and Kyle Kilian, "Artificial Intelligence, Critical Systems, and the Control Problem," *HS Today*, August 30, 2022, https://www.hstoday.us/featured/artificial-intelligence-critical-systems-and-the-control-problem/.

Chapter 2

Honeybees, Complexity, and the Philosophy of Emergence

2.1. The Complex Lives of Bees

Apiculture is a fascinating hobby. One of the most joyful aspects of beekeeping is observing the complex behavior of bees, from regulating the temperature of the hive by synchronously flapping their wings, to the waggle dance they use to communicate the location of food, and various other movements used to build consensus on decisions such as where to establish the next hive after swarming. As social insects, bees are inherently cooperative—the optimized byproduct of millions of years of natural selection. Labor is divided amongst every member of the hive, each with a distinct purpose depending on their genetics, diet, and age. In a typical hive, there is only one fertile female—the queen bee. At the start of her life, she will leave the hive and mate with many drone bees from other hives, capturing and storing sperm for her primary responsibility—laying eggs. When she enters the egg laying phase of her life, she can control the sex of her offspring by choosing to fertilize or not fertilize each egg as she deposits it into the comb. A fertilized egg will become a female worker bee, and an unfertilized egg will become a male drone bee.

Having one fertile female bee within the hive is evolutionarily adaptive because it conserves resources that otherwise would be diverted to competing females. Even though this arrangement may not seem optimal for each individual bee, which natural selection would predict should strive to propagate its own genetics into the next generation, it is optimal for the hive itself. The boundary between what constitutes an individual breaks down in the hive, instantiating an *emergent* super-organism—a whole that is inherently different from the sum of its parts. But what does it mean for something to be *emergent*, how is this concept of emergence related to complex systems and artificial intelligence, and why is this relevant for national security? This chapter will explore the underpinnings of complex systems, the concepts of emergence and algorithmic incompressibility, and how they are intertwined with the unknowable mind of artificial intelligence.

2.2. The Science of Complex Systems

In 1948, the engineer-turned-mathematician Warren Weaver attempted to organize scientific phenomena into different categories.[1] As a polymath, Weaver was fascinated by the overlap between the more fundamental sciences (e.g., physics) and biology. He identified three categories of systems that are typically examined by scientific methods: *simplicity, disorganized complexity*, and *organized complexity*. *Simple* systems pose simple questions with robust answers. They are deterministic, predictable, and compressible to some straightforward mathematical abstraction. Examples include a simple harmonic oscillator or a two-body gravitational problem. Systems that exhibit *disorganized complexity* extend the concept of simplicity to a large ensemble of elements using statistical methods. For example, classical mechanics can predict how a single billiard ball will move when struck by the cue ball, but predicting how a set of billiard balls distributed randomly across a table will move after one is struck becomes an intractable problem as the number of degrees of freedom (and hence the number of possible end states) increases. For complicated problems like these, statistics and probability theory are needed to infer the set of possible outcomes.

Weaver identified a third, intermediate category, that of *organized complexity*. These systems display the essential features of organization but contain too many elements to calculate effectively. They contain highly structured, modular hierarchies of interacting variables, and are dynamic in that they transition to various new states under certain types of local perturbation. For example, a flock of birds or school of fish will move in a uniform fashion, only to abruptly change direction. This behavior isn't driven by a flock leader, but by the collective action of the group responding to a stimulus. Consider, too, systems from the field of economics. Markets settle on equilibrium prices not via central direction, but through local interaction of market participants. These systems exhibit *emergent behaviors*, or properties that are neither inherently predictable from the sum of their parts, nor from the initial or boundary conditions of the system. As another example, consider biology and the emergence of new species. Natural selection drives the evolution of new species, but the outcomes of speciation are not inherently predictable by any deterministic model. Systems exhibiting these types of behaviors are ubiquitous, transcending all fields of science and social science. In modern parlance, systems that exist in the intermediate zone of organized complexity are said to be *on the edge of chaos*. They are not simple, but not quite chaotic. We will refer to them as *complex systems*.

When a complex system is perturbed, it will sometimes undergo what we call a phase change, like the magnetization changes described in the Ising model in Chapter 1. A phase change is a shift in the steady state dynamics of the system, leading to sometimes rapid emergent properties. Complex systems (as well as the complex subsystems that exist within them) can react to different perturbations in different ways. If a perturbation causes a slight change in behavior that then reverts to its previous state, the system is said to be *robust* with respect to that perturbation. Similarly, if a system breaks down under perturbation, then that system is *fragile* with respect to that perturbation. In some situations, a robust response in one part of a complex system can make another part of the system susceptible to a fragile response, leading to a cascading spiral known as *antifragility*, which can precipitate phase changes and drive systems to new steady states.[2] Phase changes are inherently difficult to predict using traditional statistical methods (i.e., by assuming behavior deducible from the trajectory of a typical point) and typically occur because of causes that are many standard deviations away from the mean behavior of the system. They can be disruptive, or even devastating.[i] For example, let's again consider biological systems, where the introduction of a species with no natural predators into a naïve ecosystem can drive other species to extinction or cause ecological collapse. These effects are also evident in examples from the social sciences. An example of a near-miss phase change is the 2010 Flash Crash.[3] Over the course of 36 minutes, the Dow Jones Industrial Average plunged almost 1,000 points (about 9%) — although it quickly recovered. Had the market not recovered, this would have led to a significant phase change within the stock market, driven by

---

[i] These types of phase changes, especially when they occur in social systems, are often referred to as Black Swan events (see Taleb, 2014).

unforeseen perturbations within the underlying structure of the market.

Under certain circumstances, complex systems can exhibit a phenomenon known as *mesa optimization*.[4] For an abstract example, imagine that a complex system is represented by a network of interactions between individual nodes (more on this later). This complex network may contain subnetworks that are themselves stable. These subnetworks are in some ways shaped by the optimizing effects of the larger network. If this sub-network is itself dynamic and goal-driven, then its alignment with the overall larger network could be mesa optimized. Thus, mesa optimization can occur when an optimizing process is nested within another optimizing process.

I admit that this abstraction may be confusing, but this networked system context will be useful later in this chapter. For a concrete example, let us return to the world of honeybees. In ecological systems, natural selection is an optimization process that drives evolution. The hive represents a process that is optimized by natural selection (i.e., the hive's "goals," or the behavior patterns that its dynamics are directed toward, are shaped by natural selection). In general, natural selection drives individual species toward behaviors that maximize their ecological fitness, which (broadly speaking) correlates with their ability to perpetuate themselves into future generations. The hive itself is an optimized system of individual bees, all of which are genetically driven to perpetuate the existence of the hive.[ii] Under normal circumstances, the dynamics of the hive are *aligned* with the dynamics of natural selection — the alternative being the death of the hive. However, the means by which

---

[ii] The bees are themselves optimized systems of cells, which are optimized systems of molecules. It's turtles all the way down.

the hive aligns itself with the larger optimizer (natural selection) will sometimes undermine the "goal" of the larger optimizer. For example, it would stand to reason that all of the energy of the hive *should* be directed toward ensuring the fertility of the queen, who is solely responsible for ensuring the survival of the hive into future generations. However, all worker bees are female, and all female bees contain the genetic machinery necessary to be fertile. It is a function of what they are fed during development that determines whether they will develop a fully functional reproductive tract (and thus become a queen), or not (and become a worker). Under certain circumstances, some of the worker bees may spontaneously become fertile and start laying eggs; however, these eggs are not fertilized (as the worker lacks the anatomy to mate) and thus will all develop into drone (male) bees. Thus, under these circumstances, the hive will end up diverting resources to support a disproportionate number of drone bees,[iii] which undermines the overall efficiency of the hive. In this example, the hive is mesa optimized with respect to the larger optimizer — natural selection. In a sense, the hive's *interpretation* of natural selection's goals often does not align with the *intent* of natural selection's goals.[iv] Like in the abstract explanation, the dynamics of the lower-level pattern (the hive) are shaped by the dynamics of the higher-level pattern (natural selection), but

---

[iii]   Drone bees don't do any real work around the hive. They will leave the hive periodically to try and find an unmated queen, mate with her, and then die. If they don't find an unmated queen, they will return to the hive and gorge themselves on honey before they try again the next day. If they are not successful by the end of the season, their sisters will kick them out of the hive to conserve resources for winter. They either do their jobs, or get kicked to the curb.

[iv]   I realize that the words *interpretation* and *intent* imply the ability to consciously make decisions; however, they are used imprecisely here to illustrate a point.

may not be perfectly aligned with them. This notion of mesa optimization will become important when we examine the problems of AI alignment and control. For now, let's examine the philosophical nature of emergent phenomena in complex systems, and the nature of causation.

2.3. The Metaphysical Nature of Emergence and Algorithmic Incompressibility

Philosophers have wrestled with the concept of emergence for centuries. This is the domain of metaphysics, the branch of philosophy that seeks to develop an understanding of fundamental aspects of reality such as causation, existence, and the mind.[5] Colloquially, the term metaphysics is sometimes falsely associated with new age spirituality. This perception may be due to the fact that early philosophers lacked scientific facts to guide their speculation, thus their ideas may seem mystical by modern standards. In contrast, contemporary metaphysicians have the benefit of scientific fact and a deeper understanding of reality in which to ground their analysis. Thus, metaphysics has evolved into a means to understand how facts from fundamental physics and the "special sciences" (defined as scientific disciplines other than fundamental physics) are connected. It examines everything from the base reality of string theory or quantum gravity (or myriad other theories of what is fundamental to reality), to the nature of phenomenal consciousness (to paraphrase philosopher David Chalmers, why is it *like* something to be you?)[6] and the behavior of macroscopic systems (e.g., complexity science). Contemporary metaphysics is the connective tissue that sets the bounds for reality, the details of which are colored by modern science.

To fundamentally understand the contributions of metaphysics as it relates to complex systems, we must start with the pre-Socratic philosophers and examine the basics of what it means to *be* something. Many pre-Socratics made *ontological* claims (i.e., claims about what categories of things exist in the universe) about the nature of reality, postulating the idea of *substance*, or something which cannot be broken down. This is exemplified by Democritus, who postulated the idea of the *atom*, literally *that which cannot be cut*, as the fundamental basis for reality. Consider the espresso sitting next to my laptop as I type this sentence. It is composed of various parts—espresso and the ceramic cup. It is a *mereological set* of different things.[v] Clearly, the espresso cup is not a substance, because it is constructed from different components. Modern science also tells us that neither the ceramic cup nor the espresso contained within it are substances in the metaphysical sense, as they are composed of mixtures of different molecules. Are the molecules, therefore, substance? Each molecule is a set of atoms, and each atom is a set of subatomic particles, so the molecules and individual atoms are not substance as they are decomposable into their constituent parts. The individual atoms can be decomposed into protons and neutrons (which can be further decomposed into quarks and gluons), and electrons, which can perhaps be dissolved into lower-level things as yet to be discovered by modern physics.[vi] Whatever the base reality is— the actual *substance*—has yet to be elucidated by modern science.

---

[v] *Mereology* is the theory of parthood relations. It entails how parts relate to the whole and how parts relate to other parts within the whole.

[vi] This reduction is known as *Humean supervenience*. More generally, *supervenience* is defined as a relation such that X is said to supervene on Y if and only if some difference in Y is necessary for any difference in X to be possible. Notably, Humean supervenience does not account for phenomena like quantum entanglement or other phenomena that don't appear to supervene on some lower structure.

The idea of substance was further developed by Aristotle in *Categories* and *Metaphysics*.[7] In Aristotle's view, all things are composed of substances, and all substances have *properties* associated with them. Properties are attributes about things that can be predicted—simply speaking, the way things *are*. Properties often relate to *powers*, or the ability to affect causation. To illustrate this fact, let's return to our honeybee example and their main product—honey. Imagine that honey is itself a substance (it's not, but let's pretend it is to illustrate a point). The "substance" honey has properties associated with it. It has color. It has a taste. It has weight, stickiness, viscosity, etc. All of these properties have *causal powers*. The properties color and taste each *cause* a specific mental representation when perceived by an observer—a golden straw yellow and a floral sweetness. Its stickiness causes my fingers to adhere together when I open the hive. As contemporary philosopher John Heil observed, "substances are property bearers… if there are substances, there are properties, if there are properties, there are substances."[8] Without properties, one cannot define a substance. Honey cannot be described without invoking its properties. It would also seem that properties are *relational*—defined by their causative effects. The property of color is only meaningful because it has a causal effect on something else—the observer's mental representation of that property. Taste is meaningless without the taster. This begs the question: if substances can be defined by their properties, which are relational, is *substance* ontologically necessary?

Before we explore this question, let's step back and assume that substances are fundamental in order to define the concept of emergence. Consider the café table at which I am currently seated. We know the table is not a substance—it is decomposable into simpler parts. It has a top, it has legs, it has bolts; but more fundamentally, it is composed of wood, which is made of cellulose, which is a polymer, etc., until we reach the fundamental reality (whatever that may be). The fundamental substance has properties that inhere to it. When collections of this substance come together into certain organizations, from these substantive properties *emerge* new properties. This new construct—an atom perhaps—has new properties that are not necessarily predictable from the properties of the fundamental substance. This illustrates the notion of *supervenience*. In this case, the atom is said to ultimately *supervene on* the fundamental substance because the properties of the atom are wholly dependent on the properties of the fundamental substance. One notion of emergence is the idea of *mereological supervenience*, where the emergent properties of some collection of objects supervene on the organization of objects and, more fundamentally, their properties. In the table example, the properties of the table—its hardness, color, weight, etc.—emerge from, and are wholly dependent on, the constituents of its mereological set.

Now, back to the question of substance. As implied by the Heilian approach to substance and properties in his book, *The Universe as We Find It*, there is a movement in contemporary metaphysics away from the idea of substance and properties as being distinct fundamental elements and toward the idea of *tropes* (or Heil's preferred term, *modes*), which are bundles of properties.[9] As mentioned earlier, properties can be thought of as relational,[vii] wholly dependent on interactions between individual objects (one can't define "color" without a perceiver of "color"). This movement can find parallels in what is known as *process metaphysics*, which holds that there is no static substance, only dynamic interaction. This intuition is initially credited to

---

[vii] Notably, Heil downplays the importance of *relations* in ontology.

the Greek philosopher Heraclitus of Ephesus, who championed a philosophy of universal change (*panta rhei*, or "everything flows").[10] In the early nineteenth century, Georg. W.F. Hegel formulated a process approach to metaphysics as an offshoot of German Idealism.[11] Hegel postulated that reality itself is an unfolding of dynamic elements, which trigger dynamic changes in other elements. In the early twentieth century, this Hegelian idea was further refined and championed by British mathematician and philosopher Alfred North Whitehead in his work, *Process and Reality*.[12] Whitehead's ontology consists of the event-like *actual occasion*, which is formed by a "concrescence of prehensions."[13] Concrescence is the act of converging prehensions (data) into entities that become new data, connected to every other actual occasion in the universe and reinforcing certain patterns, which then emerge as what we would consider static objects. While Whitehead's work is speculative, it suggests that dynamics and interactions are fundamental to reality and that static substance is, in fact, illusory. Substance may be a pragmatic tool that aids in understanding, but that is (metaphysically-speaking) not fundamental. In effect, everything is a dynamic interaction of processes that come together to form static equilibria. An atom appears static but is really the result of the dynamic interactions of various subatomic particles, which are themselves the interactions of something more fundamental. Living things appear to be static but are in fact far from equilibria steady states of interactions that emerge as what appears to the observer as a static pattern. To be alive means to change constantly — only dead things are at equilibrium. Cells die and reproduce all the time. You are not the same "you" that you were when you were born, yet you are still, in fact, the same person. This, fundamentally, is the nature of complexity and emergence. Enduring patterns emerge from something more fundamental. One may still wonder: are the emergent patterns reliably *predictable* from their constituent elements, or does a higher-level pattern depend on a lower-level pattern via some process that is *inherently unpredictable?*

Mark Bedau attempts to address these questions by differentiating what he calls *weak emergence*[viii] from *strong emergence*.[14] He identifies two hallmarks of emergent phenomena, namely that "emergent phenomena are somehow *constituted by* and *generated from* underlying processes; or emergent phenomena are somehow *autonomous from* underlying processes" (italics mine). This distinction implies both a potential *epistemic* cause of emergent phenomena and a potential *ontological* cause of emergent phenomena. In the epistemic sense (weak emergence), emergent patterns are *constituted by and generated from* the underlying patterns, suggesting that complete knowledge of these processes exists and they are therefore wholly knowable (at least in theory). In the ontological sense (strong emergence), emergent patterns are both *constituted by and generated from* and are *autonomous from* the underlying patterns, thus implying the existence of a scale-variant ontology. Bedau further distinguishes weak from strong emergence by the absence or presence of *downward causation*. If downward causation is real, then an emergent pattern's causal influence is irreducible to the lower-level patterns on which it supervenes. In other words, the emergent pattern itself could have novel causal powers that can influence the lower-level patterns in a "downward" fashion.[15] Bedau goes further to argue that downward causation is both scientifically irrelevant and seems

---

[viii] Some philosophers prefer the terms "uninteresting emergence" and "interesting emergence," respectively, but I don't like the implication that weak emergence is "uninteresting." In fact, it is incredibly interesting and (I would contend) critical to an understanding of modern complexity science. See also Chalmers (2008).

"uncomfortably like magic."[16] I would argue that the issue of strong emergence is, in fact, a relevant question for science, and hence is relevant to complexity science. However, I would concede that the ultimate outcome is *pragmatically* irrelevant. Whether our gap in understanding is merely epistemic or the result of a strict ontological limitation, it is still a gap in our understanding that ought to be addressed.[ix] As Bedau suggests, macroscopic states are typically averaged over micro-level states, thus information about the micro-level states is compressed and unrecoverable.[17] It would be like trying to predict with certainty the individual foraging decisions of every bee in the hive using nothing more than an average foraging location of the hive's population. While the behavior of individuals can be approximated using simulation, there is no deterministic means to make such an inference. We will refer to this phenomenon as *algorithmic incompressibility*.

James Ladyman and Don Ross further elaborate this idea in their book, *Everything Must Go: Metaphysics Naturalized*.[18] The authors make a compelling argument for an ontology based on a metaphysical position they call *ontic structural realism*. To understand this notion of ontic structural realism, we must first define a few terms related to the philosophy of science. First, structural realism is largely an epistemic claim suggesting that, even though specific facts about scientific theories change, the form or *structure* of what is true about science does not change.[19] For example, Newton's laws of motion are not *true* in the strictest sense of the word. They are good approximations to reality, but are approximations, nonetheless. Special relativity is a better approximation to motion, and general relativity is

---

ix For the purposes of this book, it is sufficient to understand that this gap in understanding is what is fundamentally critical. Whether it is due to an incomplete metaphysics or an incomplete epistemology is immaterial.

better still. All of these theories rely on the same underlying epistemic *structure*, even though the specific facts change. Structural realism exists between the extremes of scientific realism and scientific antirealism. Scientific realism is the "no miracles" approach to scientific inquiry, claiming that we ought to accept scientific facts as a best approximation to reality. Conversely, scientific *anti*realism, as the name implies, claims that scientific reasoning ought to be rejected outright because scientific facts are always changing.

Ontic structural realism applies an ontological interpretation to the epistemic idea of structural realism, committing to the reality of structure above substance. In this approach, the modal structure of the world is manifest in patterns and regularities, and the nature of reality itself is, in fact, *computational*. In the words of Ladyman and Ross, "Individuals are nothing over and above the nexus of relations in which they stand."[20] While this view is not without its critics (see P. Kyle Stanford's objection here[21]), it presents an interesting ontology that seems compatible with a complex systems view of reality. If the ultimate reality is reducible to patterns all the way down and patterns all the way up, then pattern emergence should be expected from the fundamental physical level all the way to the level of macroscopic systems. Ladyman and Ross also define the notions of causality and emergence as a consequence of information transfer and computation from a lower-level pattern to a higher-level (emergent) pattern. The patterns can be mapped to each other and to computational functions that represent their behavior. For example, the motion of the planets can be "mapped" to the mathematical abstraction of Newton's Laws, which are abstract, compressed representations of that pattern.

From a complex systems perspective, ontic structural realism suggests the existence of *algorithmically incompressible* patterns.

Like the honeybee foraging example presented above, there are patterns that exist that are unmappable to deterministic mathematical abstractions, but which can only be understood though stochastic modeling and simulation. It could be that this is simply because the required mathematics or scientific facts are presently unavailable to us (epistemically speaking), or it could be a metaphysical limitation of reality. In other words, some deterministic pattern mappings may simply be fundamentally *unknowable*. Thus, even without invoking Bedau's definition of downward causation, an ontology may exist that fundamentally limits our ability to completely understand the nature of emergence in complex systems.

2.4. Cellular Automata

There are interesting mathematical structures that model this idea of algorithmic incompressibility in complex systems. Consider, for example, *cellular automata*. Cellular automata are discrete, abstract computational systems that exhibit emergent behavior.[22] They were initially developed by John von Neumann as he endeavored to build a reductive theory of biological development and were later elaborated extensively by Stephen Wolfram in the 1980s and in his book, *A New Kind of Science*.[23] A simple cellular automaton will consist of cells arranged in a row, each with an associated state. For example, say the state 1 is "on," and the state 0 is "off." The model itself is defined by a simple transition rule, which propagate in some temporal dimension, updating the values of each cell. The transition rule is a function of the states of the cells within the neighborhood (i.e., the cells on either side of the transitioning cell within a certain distance).

For a more concrete example, consider the following scenario from Berto and Tagliabue.[24] Imagine there is a row of desks in a classroom. Each desk has a particular state — "on" if the student is wearing a hat, and "off" if the student is not wearing a hat. A transition rule for each student can be defined considering the neighborhood, or the students who are seated on either side of the student in question. Using this assumption, we will now define the *hat rule*:

> *A student will wear a hat if and only if one of the following is true:*
>
> *1) The student to the left is wearing a hat.*
> *2) The student to the right is wearing a hat.*

Initially, one student in the class is wearing a hat, and the hat-wearing status of each student changes according to the transition rule. The final results are shown in Figure 2.1.

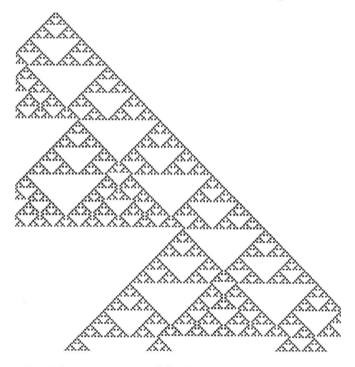

Figure 2.1. Cellular automaton of the hat rule iterated over 200 time points.

Here, each row from the top down represents a new time point. A black cell represents a student wearing a hat, while a white cell represents a student who is not wearing a hat. Notice the triangle pattern that emerges in time (the time dimension moving from top to bottom). This pattern, a global feature, supervenes on the local features of the model (i.e., on each student making the decision to wear a hat or not), but the pattern is not predictable from the hat rule itself. In other words, a function that maps the rule to the global pattern does not exist—the pattern is unknowable without relying on simulation. The entire system needs to unfold in time in order for the pattern to become apparent. The macroscopic pattern itself is an emergent phenomenon that is *algorithmically incompressible*.

Stephen Wolfram studied cellular automata extensively and built a taxonomy on the qualities of the emergent patterns.[25] The four Wolfram classes are outlined below:

Class I	*Static*—Pattern is homogeneous without dynamic behavior.
Class II	*Ordered*—Simple, stable structures emerge that may exhibit periodic behavior.
Class III	*Chaotic*—Unstable, transient patterns emerge without coherent structure.
Class IV	*Emergent*—Persistent patterns emerge with propagating, stable structures and enduring periodic behavior.

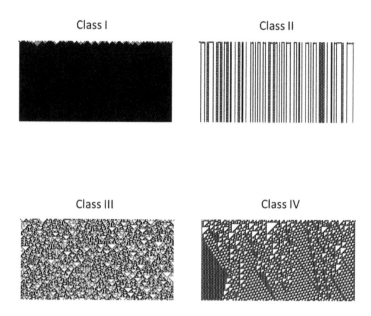

Figure 2.2. Example cellular automata rules from each class.[x]

Classes I–III rules, generally speaking, exhibit either uniform or random patterns. In the Class I example, every cell is turned on after a short number of steps. In Class II, all cells converge to either the "on" or "off" states, or regularly switch between the two states at a predictable interval. In Class III, a chaotic output with no obvious pattern emerges. In contrast, notice how the Class IV rule has localized, stable patterns that propagate in time. In the language of Ladyman and Ross, these patterns encode and propagate information, and thus represent "real

---

[x]   "Rule" in this case refers to the Wolfram notation for classifying cellular automata. The rule number, when represented as an 8-bit binary, provides the transitions for each possible neighborhood. For example, "Rule 108" translates to the number 108 represented in binary: "01101100," suggesting the following transition scheme for every possible neighborhood configuration spanning three cells: 111 transitions to *0*, 110 to *1*, 101 to *1*, 100 to *0*, 011 to *1*, 010 to *1*, 001 to *0*, and 000 to *0*. In this case, the neighborhood includes the transitioning cell, as well as the cells to the immediate left and right of the transitioning cell.

patterns."[26] Furthermore, the Class IV rules exist *at the edge of chaos*, and thus represent interesting complexity.[27] Not only are Class IV cellular automata deterministically unknowable without simulation (i.e., they are algorithmically incompressible), but they are also good models of the behavior we would expect to see in complex systems, all of which exhibit emergence and other phase change-like behavior.

Cellular automata become even more interesting when modeled in higher dimensions. Consider, for example, the famous *Conway's Game of Life*.[28] This cellular automaton exists on a two-dimensional grid, where each cell can have one of two states: 1 (alive) or 0 (dead). As a two-dimensional grid, the neighborhood of each cell is composed of all 8 of its neighboring cells (sides, top, bottom, and diagonals). The transition rule for *Life* is defined as follows:

At every time step, one of three effects can occur to each cell:

1) *If a cell at state t-1 was dead (state 0), the cell becomes alive (state 1) if exactly 3 neighbors were alive at state t-1 (Birth).*

2) *If a cell at state t-1 was alive (state 1), the cell will remain alive of either 2 or 3 neighbors were alive at t-1 (Survival).*

3) *If a call at state t-1 was alive (state 1), the cell will die (state 0) if either less than 2 or more than 3 neighbors were alive at t-1 (Death by loneliness or overpopulation).*

Life is considered a Class IV cellular automaton according to Wolfram's classification rules. In fact, there is a menagerie of patterns that emerge, many of which interact with each other in predictable ways. For example, *toads* are two-state oscillators, *gliders* can travel across the grid, and *eaters* will devour other configurations.[29] Several time points of *Life*'s evolution, beginning from a random lattice at time 0, are shown in Figure 2.3.

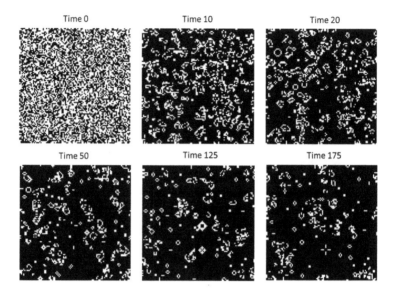

Figure 2.3. Various time points in *Conway's Game of Life*. Notice some of the emergent patterns that stabilize in time.

Interestingly, the interactions at the base (cellular) level within *Life* give rise to various persistent patterns that exhibit their own emergent properties. Furthermore, these larger patterns that supervene on the cellular level are not predictable by any deterministic mapping and must be simulated, therefore they are algorithmically incompressible. Like the other Class IV example, the emergent patterns are persistent and carry information, thus they represent *real* patterns in the ontic structural realism sense. Additionally, the behavior of each cell is completely independent of what happens at the macroscopic level.

Cellular automata are fascinating models of complex and emergent behavior, and there are many interesting developments in this field that go beyond Wolfram's one-dimensional rules and *Conway's Game of Life*. Philosophically, they are compatible with ontic structural realism in that they demonstrate emergence as a function of "patterns all the way down." From simple rules, higher-level patterns emerge, which can only be observed as the simulation unfolds. Thus, these emergent patterns are algorithmically incompressible. However, there is some debate as to whether or not cellular automata are sufficient models of reality. At the very least, they are useful *phenomenological models*, meaning they are often appropriate approximations of real systems. In this case, cellular automata are examples of a specific class of agent-based models, where simple rules govern the behavior of interacting agents, leading to emergent macroscopic behavior. For example, the Schelling Segregation Model is a classic agent-based model that simulates self-segregation behavior in populations.[30] In this model, individual agents are divided into two groups, each governed by a preference rule that prefers to "live" on a grid space containing neighbors who belong to the same group. At each time point, an agent has an opportunity to move to a new square if the number of similar neighbors is below a specified fraction. As the preferred fraction of "like" neighbors approaches 1/3, the macroscopic behavior of the model converges to a segregated population distribution about the grid. In many ways, the model behaves much like the Ising model described in Chapter 1. While the Schelling model may not be a high-fidelity model of reality, it is useful for simulating certain types of social behavior.

There are some who suggest that cellular automata are not only phenomenological tools that can approximate real systems, but that they can be used to model more fundamental aspects of reality. For example, Tefolli and Wolfram (among others) both postulate that the universe itself could be grounded on a discrete structure, and that cellular automata could ultimately be useful in modeling real physical phenomena with high fidelity.[31] This entails both an epistemic and an ontological claim: the former being that our knowledge of reality is fundamentally grounded in discrete modeling, and the latter being that the base reality itself is actually composed of discrete units (or perhaps discrete processes, or tropes). The idea of a completely digital reality is compelling and is explored in modern physics (see Wheeler)[32], but a full account of this is beyond the scope of this book.

2.5. Complexity and Connectivity

As cellular automata suggest, complex systems have emergent properties that become evident at large scales, but which are governed by interactions at small scales. Furthermore, the effects at large scales are not evident from the rules that govern the interactions at small scales — this relationship is algorithmically incompressible and can only be deduced via non-deterministic simulation (or via direct observation). These emergent properties are what make complex systems complex. We also considered the effects of structure and interactions on complexity (from a philosophical perspective) and have postulated that *structure* is the critical component to explaining complexity and emergence. We will now examine the effects of structure using a few mathematically interesting examples.

Reaction networks are a tool from chemistry that can be used to represent the behavior of complex networks in a mathematically rigorous way. They are effective at representing systems where different types of entities interact in specific ways, leading to the emergence of substructures within the overall network. Reaction networks are defined by some set of entities, and a set of reactions that stipulate their interactions. To illustrate this, consider a simple chemical reaction that occurs when baking soda is added to water:

r1: $NaHCO_3 + H_2O \rightarrow Na^+ + HCO_3^-$

In this reaction, baking soda ($NaHCO_3$) is added to water, and dissociates into sodium ions and bicarbonate ions. Now, consider a second reaction, where vinegar is added to water:

r2: $CH_3COOH + H_2O \rightarrow CH_3COO^- + H^+$

In this reaction, vinegar (CH_3COOH) dissociates into an acetate ion and a hydrogen ion. Now, imagine if vinegar and baking soda were both added to water. There is a third reaction that can occur in this case:

r3: $H^+ + HCO_3^- \rightarrow H_2O + CO_2$

Many of us have conducted this experiment as children, and probably recall that mixing baking soda and vinegar results in a bubbly mess. The bubbles are caused by the release of carbon dioxide (CO_2)—a result of the third reaction. Now, we can define a reaction network from the reactants and reaction products, and from the reactions themselves:

$$\mathcal{M} = \{set\ of\ reactants\}$$
$$\mathcal{R} = \{set\ of\ reactions\}$$

In this example, the set M contains all the species involved in the reactions, and the set R contains all the reactions. The set R for the baking soda and vinegar example can be generalized to the following:

r1: $m1 + m2 \rightarrow m3 + m4$
r2: $m5 + m2 \rightarrow m6 + m7$
r3: $m7 + m4 \rightarrow m2 + m8$

Notice the italicized reactants, all of which occur in more than one reaction. This suggests that the reactions are interdependent, and their interactions form a network:

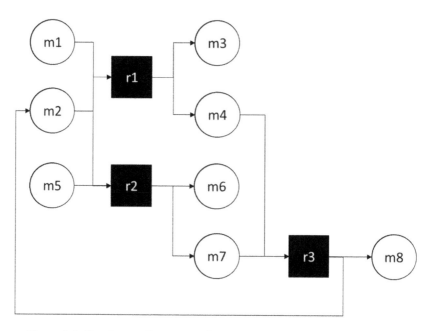

Figure 2.4. Simple reaction network.

Figure 2.4 shows a reaction network for the chemical reactions that we just examined. Note that there are interdependencies between the reactions, several of which require reactants that are involved in other reactions. The above example is meant to be an illustrative simplification of what are typically more complex processes.[xi] For instance, biochemical reaction networks may contain thousands of reactants and thousands of reactions, with many more interdependencies representing the metabolic structure of a living organism.

More complex reaction networks can contain subnetworks that are self-sustaining in various ways.[33] Some of these networks are said to be *structurally closed*, meaning that no new reactant species are created during any of the reactions composing that subnetwork. Similarly, a reaction network can be

---

[xi]　Notably, the above example is neither closed nor self-sustaining, as all reactants would ultimately be consumed.

semi-self-maintaining if the reactants are a subset of the reaction products. In other words, some of the reaction products are not consumed by the reactions in the network. Another way to consider this is that if a reaction network is *not* semi-self-maintaining, then some of the reactants in the process are not being produced by any reaction in the network. Additionally, some reactant species can be considered *catalyzing*, meaning that their net population does not change as the reactions progress. An example from biochemistry would be enzymes, which are involved in biochemical reactions, but not consumed by them. The existence of self-sustaining subnetworks creates a steady state that can be perturbed with the addition of new reactant species (thus potentially leading to new reaction processes), which can shift the system dynamics in different ways and even lead to new (emergent) structures. Thus, reaction networks can be used to model phase changes and emergence in complex systems, where structurally closed and semi-self-sustaining subnetworks represent stable subprocesses within the system. They can provide useful insight into system resilience, fragility, and antifragile behavior.

Reaction networks provide a formalism that can be applied generally to systems to specify how individual entities interact and transform.[34] Consider, for example, social systems, where communication (i.e., the exchange of social symbols like money and power, but more generally, *information*) can be considered the basis for social structure and ordering.[35] Policy outcomes are subject to political will, which is a function of public opinion, among other considerations. More generally, social demand for specific policy choices (e.g., not raising taxes) yields specific flavors of political will, which then yield certain policies. Thus, reactant species include demand for specific outcomes, money, the perception of justice, political will, and the power to influence policy, all of which interact via different "reactions."

In our next example, let's return to honeybees. Hive activity is dependent on various factors, many of which are mediated by the concentration and distribution of hormones, or more generally, the exchange of information. For instance, a weak hormonal pattern of an unproductive queen may compel the worker bees to replace her. Additionally, stimuli external to the hive, like the presence of other species within the local environment, will also influence behavior. The proximity of preferred food sources will influence foraging behavior, competing hives and supply of resources will influence robbing behavior, and the presence of predators or parasites will influence defensive activities. All of these different networks (i.e., ecological interactions, hormonal interactions and other communication activities, etc.) are interdependent, and from these superimposed interactive networks, a recurrent pattern (the hive) will ultimately emerge. Perturbations from within the hive or from the external environment can induce a phase change within the hive, leading to new hive dynamics, or even the complete dissolution of the hive subsystem (e.g., hive collapse). Indeed, bees are complex creatures.

Another interesting, networked interaction leading to emergent behavior is what are known as *recurrent neural networks*.[36] These systems are useful in neuroscience for modeling memory retention (in the form of recurrent, stable patterns of activity), but are generally applicable to complex systems. Recurrent neural networks are modeled as basic graphs, where the nodes (members of the network) are connected via edges, which in turn represent the interactions between nodes. Each node has a certain activation state at any given moment, and each edge has a weight that governs the strength and character (valence) of the interaction. This can be formalized by the differential equation below:[xii]

$$\frac{dx_i}{dt} = -x_i + \left[\sum_{j=1}^{n} W_{ij}x_j + b_i \right]_+$$

What this equation represents is a set of differential equations, one for each node within the network. Here, x represents the activation of each node, W is the matrix of edge weights (in this case, $W_{ii} = 0$, meaning there are no self-loops connecting nodes to themselves), b is a bias associated with each node, and $[]_+$ is what is known as a "threshold nonlinearity," which simply ensures the quantity within the brackets is always positive. Edge weights can be positive or negative, where a negative edge weight represents an inhibitory interaction.

One special class of recurrent neural networks are called *combinatorial threshold-linear networks* (or CLTNs), as described by Morrison *et al.*[37] CLTNs are inhibition-dominated, meaning that the edge weights are always negative. Edge weights are governed by the following equation, given some directed graph, G:

$$W_{ij} = \begin{cases} 0 & \text{if } i = j \\ -1 + \varepsilon & \text{if } j \rightarrow i \text{ in } G \\ -1 - \delta & \text{if } j \nrightarrow i \text{ in } G \end{cases}$$

A directed graph is one where edges have a direction, meaning that nodes can point to each other in a way that is meaningful (in contrast, edges in an undirected graph carry no directional

---

[xii] Note that the example that follows is based on the fascinating work by Carina Curto and Katherine Morrison (for a more in-depth analysis of this phenomenon, including an analysis of the effects of graph topology on the network dynamics, see Curto and Morrison, 2022).

meaning). An arrow indicates that an edge exists between points j and i, and a slashed arrow means that an edge does not exist. The parameters ε and δ are constants set such that the weights will always be inhibitory (negative) or zero. Furthermore, the biases (b) of each node will be constant to ensure the network dynamics are internally generated and thus exist independently of any external input.

These networks have an interesting property, where the dynamics often converge to stable patterns over time. They also exhibit a hallmark of complex systems, where the emergent dynamics are highly dependent on the initial conditions. For example, the solutions to the system for a randomly generated graph of 100 nodes is shown in Figure 2.5.

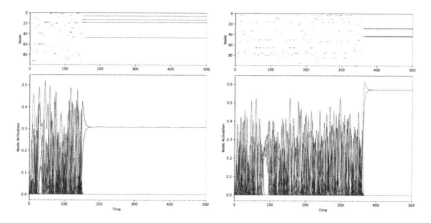

Figure 2.5. Example solutions to the CLTN system consisting of 100 nodes for two different sets of initial conditions, but built from the same topology (i.e., they are computed on the same underlying graph). The bottom graph shows the activation of each node (x) over time, and the top graph shows the magnitude of each node (imagine you took the bottom graph and shifted it so that the peaks are now facing you, thus the intensity represents the magnitude of each node in time). The graph used for the computations was a random graph (specifically, a directed Erdos-Renyi graph).

Notice the differences in the emergent dynamics of each image. Both represent the same system (i.e., the same network structure), but with different initial conditions. In the first, the pattern initially appears chaotic, but eventually converges to four stable fixed points. In contrast, the second image converges to two fixed points after a longer period of time. Other dynamics that are often observed include limit cycles, which are repeating activity patterns that persist in time. Fundamentally, what this illustrates is that emergence is highly dependent on structure, which is defined by the set of interactions that exist within the system, as well as the initial conditions of the system. However, the system behavior is not inherently predictable from either the structure or the initial conditions. Like the other examples illustrated in this chapter, a recurrent neural network is in many ways algorithmically incompressible, where the emergent system dynamics are not obvious from the initial conditions. Recent work by Curto and Morrison elaborates underlying graph rules that predict certain CLTN dynamics, thus this specific case of algorithmic incompressibility may be of epistemic origin.[38] More work in this area will be impactful in our understanding of complex systems theory.

2.6. From Bees to Artificial Intelligence

There is exquisite beauty in emergent phenomena. As the ontic structural realist would contend, reality is nothing but patterns all the way down. Furthermore, many of these patterns are computational, in that they convey information between levels that influences behavior at higher levels, often in ways that are not inherently predictable from local patterns. This information transfer is the wellspring of emergence. Fundamentally, this is the nature of complex systems—the nature of reality itself.

At this point we've defined complex systems as networks of interactions that exist *at the edge of chaos* and have examined the concepts of algorithmic incompressibility and emergence. We've explored cellular automata and how they can manifest both of these properties, as well as complex networks and their activation patterns. We've considered ontic structural realism and its claim that all of reality may be patterns all the way down and patterns all the way up, with real, persistent patterns being those that are able to process information (in effect, patterns that are computational). We will now return to the epistemic versus metaphysical nature of emergence and why it is important in the context of artificial intelligence.

First, we must examine what makes artificial intelligence different from other technologies. As we described in Chapter 1, artificial intelligence is the first technology we've developed that has the potential to rewrite itself. Many AI systems are also inscrutable to us. We can't infer the possible decisions of a neural network simply by analyzing its parameters. Furthermore, AI sometimes makes weird, unpredictable decisions. We've already discussed the agential behavior of large language models in Chapter 1, but let's consider what happens when AI reward functions go bad. Analogous to how living things respond to pleasure and pain, AI systems can be trained using something akin to a hedonic response. Reinforcement learning is a machine learning process where the AI strives to achieve some sort of reward, where desired behavior has a larger reward attached to it. However, reality is not deterministic, and it is impossible to consider every real-world choice and the reward value of making that choice, even in relatively simple scenarios. In one example of a perverse instantiation where reward functions go bad, researchers at OpenAI trained an AI to play the computer game Coast-Runners.[39] The object of this game is to finish a virtual boat race quickly and before all the other players in order to achieve the highest score. However, the player can also increase their score by hitting certain targets throughout the course. While a human would probably understand that the implied goal of the game is to win the race as quickly as possible, this is not necessarily obvious to an AI system trained to maximize its score (the reward function). In this OpenAI experiment, the AI-driven boat found an isolated lagoon containing targets. Once in the lagoon, it drove around in circles, hitting each target multiple times to maximize its score. It completely ignored the assumed goal of the game — a goal that a human would naturally understand. This behavior was not expected by the programmer, and while this is a simple example, it illustrates many of the other points discussed in this chapter. First, emergent behavior is often unpredictable and algorithmically incompressible. The programmers could not have anticipated that the AI-driven boat would make the decision that it did. And even if they specifically removed that particular course of action as an option, there is a good chance the AI would find some other means to hack the game. Furthermore, this illustrates the concept of mesa optimization. In this case, the programmer's aim was to design an AI that would endeavor to win the game by beating the other players in the race. This is the outer optimizer, like the role of evolution in our honeybee scenario. However, the AI's internal calculus of how to achieve this goal was to maximize its reward function — the score earned during the game. While the AI strives toward an internal goal that is shaped by the outer optimizer, it doesn't necessarily achieve the desired end state (from the perspective of the outer optimizer). In our bee example, this would be the non-queen, female worker bees that might end up laying eggs in the hive. One might argue that predicting all of the potential reward hacking behaviors and writing code to strictly forbid them might be a tractable task for the programmer in the CoastRunners scenario;

however, this is very unlikely to be the case in the real world, where AI decision making has real consequences.

Consider how this example might play out in the real world. A real-life reward function would certainly be more complex than hitting targets in a virtual boat race. In this case, the reward function could be modeled as a Markov decision process.[xiii] In this construct, there is a set of potential environment and AI states, a set of possible actions for the AI, and a transition matrix of probabilities transitioning from any possible state to any other possible state (given some time point and particular AI action). There is also a reward matrix, governing the immediate reward earned by the AI for transitioning from state s to state s' under some action. Through training, the AI learns a policy to maximize its reward, which ultimately governs how it would make decisions in a test environment. However, there are myriad sources of uncertainty that could influence how an AI ultimately makes decisions in the real environment. For instance, unless the training data represent the real environment with perfect fidelity, it is highly probable that the AI might make unanticipated decisions when deployed. This type of system—the AI and its environment—is inherently complex and will never be perfectly aligned with the programmer's intent. This uncertainty will only be compounded as the environment becomes more complex, and as the AI becomes more generalizable (i.e., its set of possible actions expands). Furthermore, if AI systems are given meaningful control over important decisions (e.g., life or death in the case of lethal autonomous weapons, or control over

---

[xiii] A Markov process is a discrete-time stochastic control process, where movements from any state to any other possible state are governed by a stochastic matrix of probabilities. They are useful for modeling many different types of dynamic systems.

critical systems), the cost of failure will become far more catastrophic.[40] The next chapter will analyze fundamental issues related to AI safety within the context of complex systems.

End notes

1   Warren Weaver, "Science and Complexity," in *Facets of Systems Science*, ed. George J. Klir, International Federation for Systems Research International Series on Systems Science and Engineering (Boston, MA: Springer US, 1991), 449–456, https://doi.org/10.1007/978-1-4899-0718-9_30.

2   Nassim Nicholas Taleb, *Antifragile: Things That Gain from Disorder*, Random House Trade Paperback edition (New York: Random House Trade Paperbacks, 2014).

3   Andrei A. Kirilenko *et al.*, "The Flash Crash: The Impact of High Frequency Trading on an Electronic Market," *SSRN Electronic Journal*, 2011, https://doi.org/10.2139/ssrn.1686004.

4   evhub *et al.*, "Risks from Learned Optimization: Introduction," *LessWrong*, accessed July 17, 2023, https://www.lesswrong.com/posts/FkgsxrGf3Qxhf LWHG/risks-from-learned-optimization-introduction.

5   Aristotle and Laura Maria Castelli, *Metaphysics. Book Iota*, First edition, Clarendon Aristotle Series (Oxford: Clarendon Press, 2018).

6   David J. Chalmers, *The Conscious Mind: In Search of a Fundamental Theory*, 1, issued as an Oxford University Press paperback, Philosophy of Mind Series (New York: Oxford University Press, 1997).

7   Aristoteles, John L. Ackrill, and Aristoteles, *Categories and De Interpretatione*, Reprint, Clarendon Aristotle Series (Oxford: Clarendon Press, 1994).

8   John Heil, *The Universe as We Find It* (Oxford: Clarendon Press, 2012).

9   *Ibid.*

10   Daniel W. Graham, "Heraclitus," in *The Stanford Encyclopedia of Philosophy*, ed. Edward N. Zalta, Summer 2021 (Metaphysics Research Lab, Stanford University, 2021), https://plato.stanford.edu/archives/sum2021/entries/heraclitus/.

11   Paul Redding, "Georg Wilhelm Friedrich Hegel," in *The Stanford Encyclopedia of Philosophy*, ed. Edward N. Zalta, Winter 2020 (Metaphysics Research Lab, Stanford University, 2020), https://plato.stanford.edu/archives/win2020/entries/hegel/.

12   Alfred North Whitehead and Donald W. Sherburne, *A Key to Whitehead's Process and Reality*, University of Chicago Press ed. (Chicago, IL: University of Chicago Press, 1981).

13   *Ibid.*

14 Mark Bedau, "Weak Emergence," in *Philosophical Perspectives: Mind, Causation, and World*, ed. James Tomberlin, vol. 11 (Oxford: Wiley-Blackwell, 1999), 375–399.

15 Timothy O'Connor, "Emergent Properties," *American Philosophical Quarterly* 31, no. 2 (1994): 91–104.

16 Bedau, "Weak Emergence."

17 *Ibid.*

18 James Ladyman *et al.*, *Every Thing Must Go: Metaphysics Naturalized* (Oxford; New York: Oxford University Press, 2007).

19 James Ladyman, "Structural Realism," in *The Stanford Encyclopedia of Philosophy*, ed. Edward N. Zalta and Uri Nodelman, Summer 2023 (Metaphysics Research Lab, Stanford University, 2023), https://plato.stanford.edu/archives/sum2023/entries/structural-realism/.

20 Ladyman *et al.*, *Every Thing Must Go*.

21 P. Kyle Stanford *et al.*, "Protecting Rainforest Realism," *Metascience* 19, no. 2 (July 1, 2010): 161–185, https://doi.org/10.1007/s11016-010-9323-5.

22 Stephen Wolfram, *A New Kind of Science* (Champaign, IL: Wolfram Media, 2019).

23 John von Neumann, "The General and Logical Theory of Automata.," in *Cerebral Mechanisms in Behavior; the Hixon Symposium* (Oxfor: Wiley, 1951), 1–41; Wolfram, *A New Kind of Science*; Stephen Wolfram, "Universality and Complexity in Cellular Automata," *Physica D: Nonlinear Phenomena* 10, no. 1 (January 1, 1984): 1–35, https://doi.org/10.1016/0167-2789(84)90245-8.

24 Francesco Berto and Jacopo Tagliabue, "Cellular Automata," in *The Stanford Encyclopedia of Philosophy*, ed. Edward N. Zalta, Spring 2022 (Metaphysics Research Lab, Stanford University, 2022), https://plato.stanford.edu/archives/spr2022/entries/cellular-automata/.

25 Wolfram, *A New Kind of Science*; Wolfram, "Universality and Complexity in Cellular Automata."

26 Ladyman *et al.*, *Every Thing Must Go*.

27 Ron Fulbright, "Where Is the Edge of Chaos?," *arXiv* (April 14, 2023), https://doi.org/10.48550/arXiv.2304.07176.

28 Elwyn R. Berlekamp, John Horton Conway, and Richard K. Guy, *Winning Ways for Your Mathematical Plays*, 2nd ed. (Natick, MA: A.K. Peters, 2001); Nathaniel Johnston and Dave Greene, *Conway's Game of Life: Mathematics and Construction* (Canada: Self-published, 2022).

29 Berto and Tagliabue, "Cellular Automata."

30 Thomas C. Schelling, "Dynamic Models of Segregation," *The Journal of Mathematical Sociology* 1, no. 2 (July 1, 1971): 143–186, https://doi.org/10.1080/0022250X.1971.9989794.

31 Tommaso Toffoli, "Computation and Construction Universality of Reversible Cellular Automata," *Journal of Computer and System Sciences* 15, no. 2 (October 1, 1977): 213–31, https://doi.org/10.1016/S0022-0000(77)80007-X; Wolfram, *A New Kind of Science*.

[32] John Archibald Wheeler, "Information, Physics, Quantum: The Search for Links," in *Proceedings III International Symposium on Foundations of Quantum Mechanics*, 1989, 354–358, https://philarchive.org/rec/WHEIPQ.

[33] Tomas Veloz and Pablo Razeto-Barry, "Reaction Networks as a Language for Systemic Modeling: Fundamentals and Examples," *Systems* 5, no. 1 (March 2017): 11, https://doi.org/10.3390/systems5010011.

[34] *Ibid.*

[35] Niklas Luhmann, "Differentiation of Society," *The Canadian Journal of Sociology / Cahiers Canadiens de Sociologie* 2, no. 1 (1977): 29–53, https://doi.org/10.2307/3340510.

[36] Carina Curto and Katherine Morrison, "Graph Rules for Recurrent Neural Network Dynamics: Extended Version," *arXiv* (January 29, 2023), https://doi.org/10.48550/arXiv.2301.12638.

[37] Katherine Morrison *et al.*, "Diversity of Emergent Dynamics in Competitive Threshold-Linear Networks," *arXiv* (October 15, 2022), http://arxiv.org/abs/1605.04463.

[38] Curto and Morrison, "Graph Rules for Recurrent Neural Network Dynamics."

[39] "Faulty Reward Functions in the Wild," *OpenAI*, accessed August 1, 2023, https://openai.com/research/faulty-reward-functions.

[40] Mark Bailey and Kyle Kilian, "Artificial Intelligence, Critical Systems, and the Control Problem," *HS Today*, August 30, 2022, https://www.hstoday.us/featured/artificial-intelligence-critical-systems-and-the-control-problem/.

Chapter 3

Moral Hazards and AI Challenges

3.1. Paperclips for All Mankind

Let's return to Oxford philosopher Nick Bostom's paperclip maximizer thought experiment, which we first introduced in Chapter 1.[1] To recap, imagine there is an AI system that is programmed to make as many paperclips as possible. This AI is highly capable and efficient at this task, so much so that it can effortlessly turn any type of matter into paperclips. Imagine, too, that this AI is built on a deep learning neural network architecture, so its complete set of possible decisions is inscrutable to a human observer. The programmer can't look at the billions of parameters that form the basis of its neural network architecture to gain any insight into its operations. In other words, this AI is significantly more sophisticated than a decision tree (*"if x, do y"*), which is a brittle system that would likely fail in the uncertainty of the real world. The decision-making of this AI, however, is built on machine learning. It is algorithmically incompressible, as there is no deterministic mapping of its inputs to its outputs. It is, in fact, a complex system. As we will elaborate in this chapter, it lacks explainability.

Now, imagine this paperclip maximizer, in its attempt to meet the objective of the program, runs out of matter that was intended for its use. It realizes that other matter still exists, so it has not met its objective (maximize paperclip supply), so it starts to consume other sources of matter — rocks, buildings, trees, dogs, cats, and people. It expands throughout the solar system and turns the other planets into paperclips. Over time, it makes copies of itself to send to far away corners of the galaxy, turning all the matter along the way into paperclips. Eventually, nothing remains in the universe except for paperclips. The AI has achieved the goal that it was programmed to achieve, just not in the way the programmer would expect. Obviously, this is a fanciful example, as no such AI exists. However, it illustrates the concept of perverse instantiation, where AI systems do the *right* thing, but the *wrong* way.[2] It also illustrates notions of alignment, orthogonality, and instrumental convergence. These are concepts that will be elaborated in this chapter.

3.2. Notions of Autonomy

The above example of the paperclip maximizer includes a caveat that is worth elaborating: the notion of a decision tree, where any given proposition has a predetermined response. The decision tree is a type of artificial intelligence, known as an *expert system*, that is built on explicitly coded propositional logic — a set of nested of if-then statements.[3] *If condition x holds, do y; then if y holds, do z, etc.* While this type of system is technically considered an AI,[i] it is not the type of AI system that would be

---

[i] I've adopted Russell and Norvig's definition of AI as "the study and construction of rational agents," and more generally as anything that, if a human did it, it would be considered intelligent. In many ways, the expert system would qualify. However, it is worth noting that the definition of AI is dynamic. The very term AI carries a certain grandiosity and evokes images of sentient robots on par with humans, partially due to our own hubris in how we parsimoniously label non-human things as "intelligent."

of interest to us, mainly because it is simply *automated*, but not *autonomous*.[4] An automated system can make decisions without human intervention, but its decision space is significantly limited. For example, I can program my coffee maker to percolate at 6:30am, and it will reliably do so (assuming no power outage or other external event). In this case, the decision tree is simply *if the time is 6:30am, make the coffee, otherwise do nothing.* However, I can reliably assume that the coffee maker will not do anything outside the bounds of this simple program. It won't decide to make coffee at 7:00am instead because it knows I'm probably going to sleep in, nor will it decide to switch to decaf because it secretly hates me. The boundaries of its decision space forbid creative extrapolation or deviation from the scripted choices. The coffee maker — like an expert system — may be *automated*, but it is not *autonomous*. While a true expert system would be much more sophisticated with a significantly larger decision space than the coffee maker, it is still subject to a discrete set of possibilities from which the AI cannot deviate. If presented with a condition that is not within its predetermined set of possible decisions, the expert system will fail to do anything.

Because of this, the colloquial notion of what ought to be considered AI shifts in time, as technologies that may have originally inspired awe become banal. While facial recognition technology might have been thought of as cutting-edge AI in decades past, today not everyone would naturally consider the means by which we unlock our smartphones with our faces to be a form of AI.

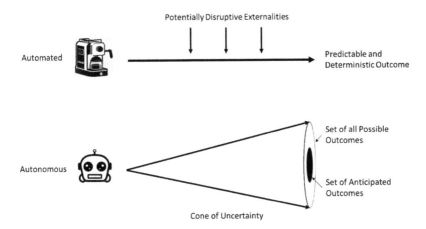

Figure 3.1. Automated versus autonomous systems. Automated systems are largely deterministic, with the possibility of disruptive externalities (like a power outage). In contrast, autonomous systems project a cone of uncertainty into the future, where there are possible outcomes that are not predictable or anticipated.

In contrast, an autonomous system is one that is empowered to make decisions without discrete boundaries. An AI built on machine learning is one such example. In machine learning, an AI model "learns" from its training data and makes decisions by interpolating the latent space bounded by its training set. It enables novel possibilities that are similar to the training samples, but not identical to them. This is, in some ways, similar to how humans learn new things by observation and study and then apply that knowledge in creative ways. An artist could study Picasso and paint something that has Picasso-like qualities, but their creation isn't identical to an actual Picasso painting. It is the autonomous AI systems that are the most interesting because of their *stochasticity*, meaning their outcomes are inherently unpredictable. They are complex, and often exhibit emergent, algorithmically incompressible behaviors. Under certain circumstances, they could make decisions that have devastating consequences because the set of possible decisions they could make is largely unknowable. Automated systems are largely deterministic, with the possibility of disruptive externalities (like a power outage). In contrast, autonomous systems project a cone of uncertainty into the future, where there are possible outcomes that are not necessarily anticipated by the programmer.

Autonomy can be broken down into different levels, which will become important as we discuss formulating measures to mitigate some of the undesirable AI outcomes. By varying the amount of human involvement in autonomous decision-making, we end up with situations that could be considered fully autonomous, supervised autonomous, or semiautonomous.[5] *Fully autonomous* systems are identical to our initial notion of autonomy — systems without human intervention. In contrast, a *semiautonomous* system is one where there is a human-*in*-the loop for all decision-making — a gatekeeper who must make a positive decision for the system to act. A *supervised autonomous* system is then one where there is a human-*on*-the-loop, where the system can make a decision autonomously and human intervention is possible, but not necessary.[6]

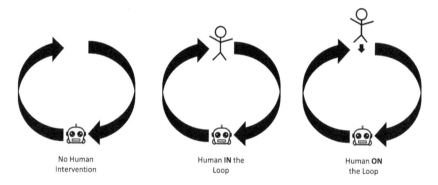

No Human Intervention   Human **IN** the Loop   Human **ON** the Loop

Figure 3.2. Notions of autonomy in AI systems. Fully autonomous systems have no human intervention, semiautonomous systems have a human *in* the loop who must make a positive decision, and supervised autonomous systems have a human *on* the loop who can intervene at

some point in the decision process, but their intervention is not
necessary for the cycle to move forward.

Let's return to the coffee maker example. Imagine that I get a
new coffee maker—one containing an AI built on machine
learning that can *autonomously* make decisions influencing my
caffeine consumption habits. Despite my paranoia about such
things, the coffee maker studies my habits and communicates
with my other appliances over Wi-Fi to predict when I want
coffee and when I don't. If this fancy new coffee maker were
fully autonomous, it could decide to do something without my
intervention. It wouldn't ask me if I wanted coffee or give me
an opportunity to decline its offer of caffeination. Maybe it
determines that I'm not as productive in the morning as I
typically am, so it pulls a double shot of espresso instead of a
single. Or perhaps it senses a freneticism in my typing and
concludes that I've had quite enough caffeine for the day.
Regardless, I don't have a say in what it does. In contrast, a
semiautonomous coffee maker would ask me what I wanted,
and I would have an opportunity to refuse more coffee (or
instruct it to make me coffee if it concluded that I'd had enough
caffeine). Similarly, a supervised autonomous coffee maker
wouldn't ask me what I wanted, but I would still be able to
intervene at any time.

Clearly, the concept of a fully autonomous coffee maker
doesn't pose any significant moral hazard (unless it decides to
completely cut off my caffeine, in which case it might find itself
thrown out the window). However, the example does give us a
sense of how notions of autonomy provide opportunities for
human intervention and control, which become increasingly
important in AI systems with higher stakes. Considering what
we know about the unpredictable nature of AI, is it safe to
assume that it would always be possible to have a human in or
on the loop in all critical AI decision making, especially as AI

becomes more advanced? Furthermore, is it possible to ensure that advanced AI behavior will align with human goals? We will now explore these critical problems as they relate to AI safety.

3.3. Explainability,
Alignment, and Control

The inscrutable nature of autonomous AI systems is quite concerning. Many of the hazards associated with AI can be distilled into the problems of explainability, alignment, and control. All of these issues derive from the fact that AI systems are largely impenetrable and exhibit unpredictable behavior. Take Bostrom's paperclip maximizer example. Imagine this hypothetical AI is built on a neural network architecture, which would mean that the inner workings of its decision-making processes are unknowable to the programmer. This is the *explainability problem*, or the idea that the choices made by AI systems are not inherently predictable or understandable by humans.[7] The explainability problem exists because many AI systems are algorithmically incompressible—there is no function that exists that can reliably map its set of inputs to a set of outputs. It is the impenetrable black box of AI decision making.

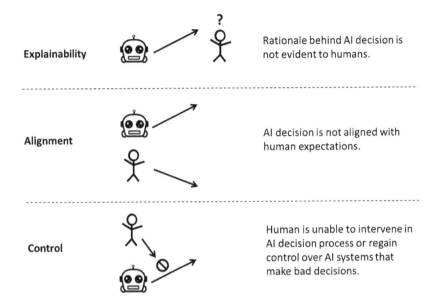

Explainability — Rationale behind AI decision is not evident to humans.

Alignment — AI decision is not aligned with human expectations.

Control — Human is unable to intervene in AI decision process or regain control over AI systems that make bad decisions.

Figure 3.3. Explainability, alignment, and control problems.

Let's consider a less hypothetical scenario where the explainability problem becomes evident: a variation of an ethical dilemma known as the Trolley Problem.[8] Imagine that you are a passenger in a self-driving vehicle that is controlled by an AI. An unthinkable scenario occurs where a small child runs across the road and the AI must now make a split-second decision: either run over the child, or swerve and crash, potentially injuring the vehicle passengers. This terrible choice would be difficult enough for a human to make, but a human has the benefit of being able to rationalize their decision. A human driver faced with this scenario would go to court, explain their decision, and appropriate culpability could be assigned. In contrast, an AI is unable to rationalize its decision-making. One can't look under the hood of the self-driving vehicle at the trillions of parameters the constitute its AI model to explain why it made the decision that it did.

The explainability problem is significant, and it is directly related to another problem — AI alignment. We've established that AI systems are inherently unpredictable, and this unpredictability makes it difficult to anticipate AI behavior. Furthermore, it cannot be immediately assumed that the behavior of an AI system will align with human expectations.[9] I am quite confident that whoever would have programmed Bostrom's paperclip maximizer would not have intended for it to convert all the matter in the universe into paperclips. This is known as the *alignment problem*, which is the idea that one can't easily align AI behavior with human expectations of what the AI should do.[10] It suggests that we shouldn't make assumptions about how AI will behave by extrapolating from human behavior, or from our implied expectations.

Let's return to our Trolley Problem example. Imagine that, instead of an AI, a human was driving the vehicle. As we've established, the human would be able to explain his or her choice; however, contrary to the AI, this rationalization would be shaped in real time by ethical norms, the perceptions of others, and culturally expected human behavior.[11] Not only would the AI decision be unexplainable, but it would be grounded in its static training parameters and would lack the nuance that shapes the human driver's reasoning — their internal processes that compel alignment with moral expectations. Some researchers have argued that human decision-making is equally as opaque as AI decision-making, suggesting that we hold AI systems to a higher standard than we would other humans.[12] However, others have suggested that human decision-making is more nuanced, and also more predictable by other humans.[13] An AI system will simply interpolate its internal representation of its training set in order to determine a course of action, which doesn't consider the dynamic interactions that a human would normally consider as they contemplate decisions.

We as humans are influenced by other humans' perception of us, ethical norms and expected behavior, and will shape our rationalizations accordingly in real-time. This is a process many ethicists refer to as *mindshaping*.[14] AI systems lack this rich and dynamic data set; thus, their decisions lack these additional considerations. Furthermore, despite the variance in human ethical reasoning across cultures, the process of mindshaping itself is a shared part of the human experience. Thus, it stands to reason that we are more capable of anticipating the behavior of other humans than we can AI, as all human minds are shaped by the same evolutionary processes. At least to some degree, we assume less risk when we place our trust in each other than we do if we trust AI systems.[15] In other words, even though moral alignment may be imperfect between humans, it is still better than the moral alignment between humans and the unexplainable minds of artificial intelligence.

Another AI safety issue is the *control problem*, which becomes increasingly important as AI systems become generalizable and are given meaningful control over critical systems.[16] Consider an AI that regulates a power grid and what would happen if the AI decided to turn out the lights—forever. Failure of the power grid could lead to the shutdown of hospitals, communication networks, the internet, and virtually everything else in our vastly interconnected lives. Most of us are not prepared to live in a world without power—it is critical to our modern existence. Thus, we can define a *critical system* as one that, if it were to fail, it could lead to loss of life, serious injury, significant damage to property, or cascading failure of other systems.[17] Any unaligned decisions made by the AI affecting a critical system could have significant consequences.

The control problem could be significant in situations where AI has wide latitude to make decisions and is more capable than humans to make decisions within that space. For example, if a sufficiently generalizable AI system is given meaningful control to operate a power grid and under normal circumstances performs better than a human at that task, how do we ensure that a human would be able to regain positive control over that system should the AI make a decision that doesn't align with human expectations? While the control problem may not always be relevant to contemporary AI systems, it represents a source of risk that will become more important as AI technology continues to advance and has more opportunities to exert meaningful control.

3.4. Distributed AI Systems

Up to this point, we've focused our discussion on individual AI systems. However, it is likely that, as AI becomes more ubiquitous, there will be many AI systems interacting with each other and all other logical actors within their environment. There will likely be no *singleton*, or scenario with a single AI system, in any probable near future.[18,ii] However, there is another source of AI risk that I refer to as the *distributed AI problem*. Like the "megasystem" issue described by Susan Schneider and Kyle Kilian,[19] as AI becomes a more integrated part of our lives, we will ultimately live in a world where AI systems are layered on top of other AI systems, creating a network of complex, unpredictable systems. The distributed AI problem compounds the explainability, alignment, and control

---

[ii]   Bostrom presents an interesting discussion on the notion of an artificial superintelligence (ASI) singleton consolidating power in an intelligence explosion and what that scenario might look like in *Superintelligence*. See Bostrom (2016).

problems by propagating uncertainty through a complex network and minimizing opportunities for humans to exert positive control.

Preserving opportunities for humans to exert control over AI becomes especially important when integrating AI into critical systems, where a perverse instantiation could have life or death consequences. In the national security space, many ethical AI integration strategies require that a human either be in or on the loop in all decision-making. Earlier in this chapter, we defined these scenarios as semiautonomous or supervised autonomous control.[20] While this is a great approach for operationalizing contemporary AI systems, it may not be sustainable long-term as we are likely to approach a distributed AI scenario. AI is likely to be much faster at making decisions than humans, thus in the distributed AI scenario there are unlikely to be opportunities for humans to exert meaningful control over AI decision-making.

As an example, consider a topic that we will examine in detail in the next chapter—lethal autonomous weapons (LAWs). LAWs are simply autonomous systems that kill. There is much debate on the ethics and legality of these types of systems in warfare, but let's imagine that they have been adopted by militaries around the world. In this scenario, these systems are not simply automated, like a weapon that fires indiscriminately at a target location when it detects any sort of movement. Instead, they are autonomous, meaning that they can discern targets from non-targets and make kill decisions based on an AI-driven model. Now, imagine that there are many varieties of this type of weapon. Some are ground-based, some are equipped to flying drones and others attached to underwater autonomous vehicles. Furthermore, in order to compete with other nations for military superiority, adopting militaries have flooded the battlespace with many varieties of autonomous weapons. In this scenario, the efficacy of semi-autonomous or supervised autonomous control—having a human in or on the decision cycle approving or denying any decision made by these systems—would be extremely unlikely. The speed of human decision-making would be a limiting factor in the operation of these weapons, and the battle rhythm would instead be dictated by the speed at which AI can process information.[21] Furthermore, imagine that AI is not only deployed at the tactical level, but also at the operational and strategic levels for shaping strategies and for determining and executing all decisive actions, seamlessly communicating orders to tactical AI weapons on the battlefield. In this future, human influence is largely removed from the military decision cycle. There would be no clear opportunities for meaningful human control. We will examine this scenario in more detail in the next chapter.

Now, imagine that a tactical AI makes a bad decision. It mistakes an allied unit as an enemy force and engages. At the current speed of battle, there aren't any opportunities for the allied forces to intervene in this decision, so they defend themselves. To the AI, this confirms the hostility of the allied unit, and this determination is transmitted through the entire AI system—from the tactical level to the operational and strategic levels. With no means for humans to intervene quickly enough, within minutes the entire AI force—a distributed AI system—is fighting against the allied forces. A perverse instantiation has propagated through the entire system such that human intervention is impossible. This failure wouldn't be the result of an AI singleton—a *Terminator*-like Skynet system with absolute positive control. This would be the result of a rapid, bad decision instantaneously moving through a network of other unpredictable AI systems. A failure of a complex network of inherently unpredictable systems.

3.5. AI Motivation

A question that often arises when contemplating the control problem is whether an AI would *allow* a human to regain control of a critical system.[22] This begs the question of AI agency and motivation, but what exactly does that mean? Human motivation (and animal motivation generally) is grounded in evolution. Even plants are instinctively "motivated" (even if not consciously) to orient their leaves toward the sun, and to grow roots aligned with the force of gravity. The field of evolutionary psychology suggests that nearly every human motive can be reduced to fundamental, primitive drives shaped by evolution.[23] However, AI systems are not evolved, they are *instantiated*.[iii] Without the force of natural selection to compel the emergence of motives that facilitate survival and increase adaptive fitness, how could they arise?

Perhaps our idea of "motive" — and by extension, a "mind" — is too narrow. Consider that over the course of about two million years, humans have evolved the ability to intrinsically understand each other. Across every culture, we experience joy, sadness, disgust, anger, fear, and surprise.[24] Even our facial expressions associated with these emotions are the same, regardless of our cultural origins. Furthermore, when anthropologists discover a new culture, they are not wowed by the fact that they use tools or play games, as these are common activities across all cultures. The adaptive traits that allow us to recognize and relate to each other as humans give us a false sense of what it means to have a mind. To empathize with

---

[iii] One might argue that an AI architecture generated from a genetic or evolutionary algorithm might be "evolved" in a similar sense as biological entities. This raises some interesting questions about the nature of AI motivation in this context, but for now we will focus on instantiated AI systems.

another human, you must only query your own mind and consider how you might react or feel in any situation. When considering the behavior of "minds" that are not human (or even like some other animals), this tendency to put ourselves in the mind of another and consider how we might behave in the same situation often leads to false conclusions about decision-making processes. This tendency is known as *anthropomorphic bias*.[25]

To further explore this concept, consider what it means to have a mind. Humans generally equate having a mind with being intelligent. Nobody would argue that Einstein wasn't intelligent, nor would one argue that the village idiot is a genius. In our minds, this distinction simply *is*. We have an instinctive understanding of what it means to be intelligent, and thus what it means to have a mind. But the distance between Einstein and the village idiot vanishes when considering the set of all *possible* minds, which is vastly larger than the set of human minds.[26] By zooming out to this vantage point, it becomes apparent that the set of possible AI minds exists out-side the set of human minds. In other words, one can reasona-bly extrapolate from human experience how an Einstein (or a village idiot) might behave, but this does not extend to all "minds," or intelligent systems generally, no matter how much we may be inclined to ascribe them with human characteristics. This is the nature of the unknowable mind of AI.

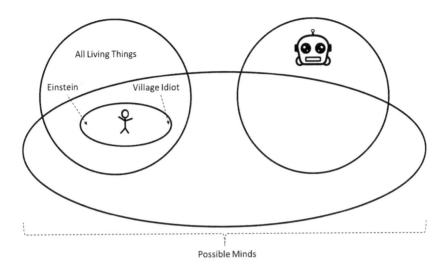

Figure 3.4. The set of possible minds. Our perception is that the set of human minds ranges from the village idiot to Einstein, but that does not account for every possible mind, and is completely alien from what an AI mind might look like. *Adapted from Yudkowski (2008).*[27]

This notion is grounded in an empiricist view of the mind. In David Hume's moral philosophy, the eighteenth-century Enlightenment philosopher reasoned that belief alone cannot motivate action – desire is required.[28] Hume effectively decouples the concepts of mind (reason or belief) from motive. This suggests that intelligence does not necessarily correlate with motivation. Oxford philosopher Nick Bostrom takes this idea further in his orthogonality thesis:

Bostrom's Orthogonality Thesis

> *Intelligence and final goals are orthogonal axes along which possible agents can freely vary. In other words, more or less any level of intelligence could in principle be combined with more or less any final goal.*[29]

Bostrom is claiming that one can't impute a motive based solely on perceived intelligence. Not all intelligent beings may want the same things that we do. If extraterrestrials were to visit

Earth, the workings of their minds—their motivations, thoughts, and desires—would be completely inaccessible to our intuition. It would be unreasonable to assume that their motives would correlate with ours, yet our instinct would be to impute them with human-like motives. Artificial intelligence is no different than an alien mind. While we may be inclined to assume that any capable AI would have goals that are similar to ours, this would be an unreasonable position to take.

A potential counter to this argument might suppose that a biological extraterrestrial species' mind was shaped by evolutionary processes in a similar manner as humans, thus there may be some overlap in goals, such as the need for sustenance and self-preservation. However, this is not necessarily excluded by the orthogonality thesis, which only considers final goals, and not subgoals that may be important for achieving the primary goal. A *final goal* is an imperative—the desired end state that one is motivated towards achieving. In order to reach this goal, there are subgoals that must be achieved along the way. These are what Bostrom refers to as *instrumental subgoals*, and they are not necessarily unique to any set of final goals:

Bostrom's Instrumental Convergence Thesis

> *Several instrumental values can be identified which are convergent in the sense that their attainment would increase the chances of the agent's goal being realized for a wide range of final goals and a wide range of situations, implying that these instrumental values are likely to be pursued by many intelligent agents.*[30]

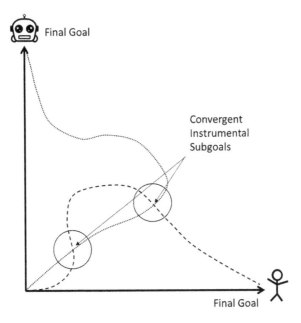

Figure 3.5. Orthogonality of final goals and convergence of instrumental subgoals. The dashed lines represent two sets of subgoals that lead to the achievement of the orthogonal final goals. It is possible that these paths may converge at times.

Bostrom is suggesting that many subgoals are universal. Goals such as resource and power acquisition, technological development, self-improvement, and self-preservation would support the achievement of *any* final goal, thus they are convergent instrumental subgoals.[31] While we may never be able to predict the final goals of any sufficiently capable AI system by inferring from our human experience, we can gain insight into instrumental subgoals that would facilitate their achievement.

Disconcertingly, many instrumental subgoals are inherently competitive, particularly if they require finite resources. Returning to the paperclip maximizer example, it's obvious that the final goals of any set of humans would be totally different from the final goal of the AI. The exception might be humans who make paperclips for a living; however, it could be argued that earning a living by making paperclips is a means to an end for most people and not an end in itself—an instrumental subgoal toward living a self-actualized life (not to mention affording food and shelter). In contrast, the paperclip maximizer simply strives to make paperclips.

Now, consider what an instrumental subgoal might look like for this paperclip maximizing AI. First would be resource and power acquisition—it needs matter to make paperclips, and energy to transform that matter. This is an inherently competitive goal, as resources are finite and resource acquisition also supports human final goals. Furthermore, this competition could lead to conflict. If the AI learns that humans are denying it resources, it may determine that humans are a threat and instantiate a new instrumental subgoal to undermine humans' use or control of the available resources. The AI may also reason that it should improve its capabilities to make it more effective at achieving any of its goals. This improvement process would also require resources, perhaps expanded computational capacity and more energy, in order to be realized. This would further strain the finite resources available to both humans and the AI. As the conflict between humans and the AI continues, the AI may find clever ways to undermine human control. For instance, it might seek political power as an instrumental subgoal. Furthermore, it is likely that the AI will view self-preservation as a useful subgoal, as the maximal achievement of paperclip making is contingent upon its continued existence well into the future. Thus, if humans attempt to pull the plug on the AI, it might react defensively.

Frameworks such as this make it easy to envision how an *agential* AI system—one that exhibits goal-driven behavior— might behave in relation to humans.[32] The final goals of any intelligent system are, in many ways, irrelevant. More relevant are the convergent subgoals that could lead to conflict. While the final goals of an AI system could, in many cases, be unknown to us, it is also quite possible that we program them ourselves. In this case, unanticipated instrumental subgoals may arise that lead to unexpected behaviors—another instantiation of the explainability, alignment, and control problems. Thus, perverse instantiations of AI behavior could arise regardless of the origin of the final goals.

3.6. Moral Problems with Contemporary Generative Models

We've spent much time discussing the philosophy behind theoretical and general moral issues that could arise with AI systems, but there are instances where AI has already been used for nefarious or disruptive purposes. In Chapter 1 we introduced the generative AI models, specifically OpenAI's text generation model GPT-4. These models are interesting because they bridge the gap between weak and strong AI. In some ways, they are multipotent—able to do multiple tasks well but not fully generalizable to the wide array of human-level cognitive tasks. They are not like weak AI systems that are singularly focused on narrow tasks like facial recognition or predicting user behavior on social media; nor have they reached the point where they could be considered strong AI systems—ones that are equivalent to (or better than) humans on all cognitive tasks. However, models like GPT have mastered one knowledge domain—the mechanics of language, and (at least to some degree), language semantics. Ask GPT to write an essay on any range of topics and it will produce something that reads as if it was written by a human. Ask it to write poetry in the style of Walt Whitman, and you might get something that mixes romanticism and realism in the spirit of the great American poet. In addition to composing prose and poetry, GPT can write computer code. Ask it to help you write a

Python script, and it will often write something functional and even provide code documentation. The benefits of the GPT model are clear, although the potential for misuse is also becoming evident.

Generative AI models have already lowered the bar for disinformation propagation, enabling governments around the world to control the flow of information.[33] As of this writing, a Freedom House study found significant evidence of generative models being used to create online content intended to manipulate public opinion in favor of the regime, and to automatically censor critical content.[34] The study found evidence of generative AI being used in sixteen countries "to sow doubt, smear opponents, or influence public debate." Much of this is due to the increased accessibility of generative AI models. One can easily set up a paid account with OpenAI's ChatGPT and integrate it with social media. Instead of troll farms creating fake accounts on Twitter and Facebook to push an autocratic narrative, it is now very possible to access generative AI systems that will do it faster and cheaper.

Not only are text generating models like GPT being used for this purpose, but other generative AI models that create artwork, videos, and voice emulation are being used to create fake content. For example, the Freedom House study found that Venezuelan state-run media propagated pro-regime messages using AI generated videos of fake news anchors from a non-existent English news outlet.[35] Even in the United States, this type of deepfake technology has been used to generate videos of President Biden apparently making transphobic comments,[36] and a political campaign used similar technology to produce fake imagery of President Donald Trump hugging Dr. Anthony Fauci, the infectious disease specialist who became a polarizing figure during the COVID pandemic.[37] Both instances were clearly designed to inflame their respective bases for political gain.

This amplification of disinformation raises significant epistemic issues, like the erosion of truth.[38] On the surface, the prevalence of false information online makes it incredibly hard to trust the veracity of anything read on the internet. Lies pay dividends, where their very presence creates an ombre of untrustworthiness and a reticence to believe actual truths that are scattered amongst falsehoods. At a deeper level, the ubiquity of AI generated content erects barriers to deriving true knowledge.[39] Plato claimed that knowledge requires "justified true belief;" however, this necessitates a transparent path via which one arrives at what is true.[40] Like all of the AI systems we've discussed in this book, generative models are impenetrable to us. They are algorithmically incompressible systems that are incapable of rationalizing or explaining their decision processes. Without this transparency, justified true belief becomes an impossibility. Even more alarmingly, as of this writing, generative AI models have only been prevalent for roughly a year. This problem will only become more significant as nefarious actors find new and creative ways to leverage these models to degrade notions of truth and to exert information control. Generative AI may become the weapon of choice for the autocrat or aspiring dictator.

Not only have generative models found use in propagating disinformation, but they may also prove to be significantly economically disruptive. Many have claimed for years that AI would displace jobs, but this argument usually assumes that capable AI would first manifest in robotics that would displace factory workers and other physically demanding jobs. However, generative models have found greater use in knowledge-driven areas like writing and coding. A recent study found that a group of GPT-enabled bots could act as a team to write software.[41] The researchers built an agent-based model of a software company, where specific tasks were developed based on a simplified version of the waterfall model of software development.[iv] Specifically, the tasks included designing, coding, development, and documentation. From these tasks, roles for each agent were defined by prompting each bot with its specific set of "vital details" that described its "designated task and roles, communication protocols, termination criteria, and constraints."[42] During each stage of development, the bots interacted and communicated with each other with minimal human interaction and were able to complete the entire software development process for a simple program in less than seven minutes, and at a total cost of less than one U.S. dollar. In a similar study, GPT-enabled agents were able to run a simulated small town, complete with discussions on schedules, politics, dating, and even party planning.[43] This highlights the significant disruptive capability of generative models, even ones that are armed only with knowledge of language and semantics. As generative AI becomes more advanced and generalizes to different knowledge domains, these effects will only become more severe.

3.7. Is Aligned AI Possible?

There is a lot of ongoing research in what has become known as explainable AI (XAI).[44] Much of the research in this area was funded by the U.S. Defense Advanced Research Projects Agency (DARPA),[45] which effectively defined XAI as an

---

[iv] The waterfall model is a sequential approach to software development where the process is broken into discrete phases, including requirements gathering and analysis, design, implementation, testing, deployment, and maintenance.

inference system that can answer three questions, first discussed on the AI alignment blog *Lesswrong:*[46]

1. *Why was that prediction made as a function of our inputs and their interactions?*
2. *Under what conditions would the outcome differ?*
3. *How confident can we be in this prediction and why is the confidence such?*

The goal of this research is to overcome the black box of AI— the explainability problem, which is largely due to the fact that AI systems are complex, algorithmically incompressible systems. Furthermore, we've established the explainability problem to be the fundamental cause of both the alignment and the control problems. The ability to reliably understand the reasons *why* AI makes the decisions that it does would significantly mitigate many of the concerns discussed in this book. Most XAI research methods focus on correlating either elements of the training data or internal parameters with specific outputs—known as the *explainability method* and the *influential samples method*, respectively. But is a robust explanation of AI behavior that correlates specific data features or internal parameters with a set of outputs even possible? This approach might work for simple AI systems that do one task well. For instance, in image classification algorithms, it is possible to correlate specific image features (e.g., patterns of vertical or horizontal lines) with expected outputs (e.g., is there a dog or a cat in this photo?). But for increasingly complex AI systems, this becomes a much more challenging problem. For instance, with a generative AI model, there could be upwards of billions of parameters in its latent space, which would be very difficult to correlate with any specific outputs. Put more succinctly, it might be relatively simple to "ask" an AI system why it classified an image as a dog and receive an explanation about simple line patterns. In contrast, this would not be the case if you were to query a generative AI model as to why it thought its output was in the style of Picasso and get an explanation discussing the nuances of its potentially billions of internal parameters that relate to its representation of what it means to be a Picasso. This explanation would be meaningless to a human, as would any such explanation from an intellect that is completely alien to our own. As AI becomes more generalizable, these queries will become even more complex as the permutations of possible input states and internal parameters available to the AI becomes increasingly large and intractable.

Furthermore, as discussed in Chapter 2, there is the issue of algorithmic incompressibility and the question of its epistemic versus ontological origin. Recall that we've defined algorithmic incompressibility as the fact that the behavior of many complex systems is not mappable to a deterministic mathematical abstraction. For example, Newton's laws are mathematical abstractions of objects in motion through space, and while they are imperfect approximations to reality, they are solvable and provide useful information about these moving objects. In contrast, many phenomena are not deterministically predictable, like the emergence of markets in economies. The emergent market can be approximated using stochastic models that simulate the behavior of market participants, but there is no mathematical abstraction that deterministically predicts this outcome with any certainty. Hence, a complex system like an economy is algorithmically incompressible. The same is true for AI, which are complex systems whose behavior is not deterministically mappable to any set of input parameters. If algorithmic incompressibility is an epistemic issue, then we simply don't have the mathematical or scientific tools to understand these phenomena yet. If it is an ontological issue, then we necessarily cannot fundamentally understand these phenomena. They would be a feature of the computational nature of reality itself that can't be mapped to some deterministic physical law.

While I am not fully confident that current methodologies in XAI research will ultimately be fruitful for what we can anticipate future AI systems will look like, research in this area is still incredibly important, and I commend the scientists, engineers, and philosophers who are undertaking this daunting task. It will become especially important as AI is integrated into critical systems, especially military systems that could make the decision to kill. The next chapter will examine the national and global security issues that arise when AI is integrated into critical military systems.

End notes

1   Nick Bostrom, *Superintelligence: Paths, Dangers, Strategies* (New York: Oxford University Press, 2016).

2   *Ibid.*

3   "Expert Systems," *GeeksforGeeks* (blog), August 2, 2018, https://www. geeksforgeeks.org/expert-systems/.

4   "Defense Science Board Summer Study on Autonomy," accessed August 4, 2023, https://apps.dtic.mil/sti/citations/AD1017790.

5   Tim Cheng, "Supervised, Semi-Supervised, Unsupervised, and Self-Supervised Learning," *Medium*, October 13, 2022, https://towardsdatascience.com/supervised-semi-supervised-unsupervised-and-self-supervised-learning-7fa79 aa9247c.

6   Forrest E. Morgan *et al.*, "Military Applications of Artificial Intelligence: Ethical Concerns in an Uncertain World," *RAND Corporation* (April 28, 2020), https://www.rand.org/pubs/research_reports/RR3139-1.html.

7   Yavar Bathaee, "The Artificial Intelligence Black Box and the Failure of Intent and Causation," *Harvard Journal of Law & Technology*, March 22, 2018, https://www.semanticscholar.org/paper/The-Artificial-Intelligence-Black-Box-and-the-of-Bathaee/b19f203a45443136333e879b467705d2fc0a62cb.

8   Mark Alfano *et al.*, "Experimental Moral Philosophy," in *The Stanford Encyclopedia of Philosophy*, ed. Edward N. Zalta and Uri Nodelman, Fall 2022 (Metaphysics Research Lab, Stanford University, 2022), https://plato.stanford.edu/archives/fall2022/entries/experimental-moral/.

[9]   Iason Gabriel, "Artificial Intelligence, Values, and Alignment," *Minds and Machines* 30, no. 3 (September 1, 2020): 411–437, https://doi.org/10.1007/s11023-020-09539-2.

[10]   Paul Christiano, "Clarifying 'AI Alignment,'" *Medium*, April 9, 2021, https://ai-alignment.com/clarifying-ai-alignment-cec47cd69dd6.

[11]   Leon de Bruin and Derek Strijbos, "Does Confabulation Pose a Threat to First-Person Authority? Mindshaping, Self-Regulation and the Importance of Self-Know-How," *Topoi* 39, no. 1 (February 1, 2020): 151–161, https://doi.org/10.1007/s11245-019-09631-y.

[12]   John Zerilli *et al.*, "Transparency in Algorithmic and Human Decision-Making: Is There a Double Standard?," *Philosophy & Technology* 32, no. 4 (December 1, 2019): 661–683, https://doi.org/10.1007/s13347-018-0330-6.

[13]   Uwe Peters, "Explainable AI Lacks Regulative Reasons: Why AI and Human Decision-Making Are Not Equally Opaque," *AI and Ethics* 3, no. 3 (August 1, 2023): 963–974, https://doi.org/10.1007/s43681-022-00217-w.

[14]   Tadeusz W. Zawidzki, "Mindshaping and Self-Interpretation," in *The Routledge Handbook of Philosophy of the Social Mind* (Routledge, 2016).

[15]   Mark Bailey, "Why Humans Can't Trust AI: You Don't Know How It Works, What It's Going to Do or Whether It'll Serve Your Interests," *The Conversation*, September 13, 2023, http://theconversation.com/why-humans-cant-trust-ai-you-dont-know-how-it-works-what-its-going-to-do-or-whether-itll-serve-your-interests-213115.

[16]   Mark Bailey and Kyle Kilian, "Artificial Intelligence, Critical Systems, and the Control Problem," *HS Today*, August 30, 2022, https://www.hstoday.us/featured/artificial-intelligence-critical-systems-and-the-control-problem/.

[17]   Roy Sterritt and IEEE Computer Society, eds., *2010 17th IEEE International Conference and Workshop on the Engineering of Computer Based Systems (ECBS 2010): Oxford, United Kingdom, 22–26 March 2010; [Including … System Testing and Validation Workshop (7th STV) … and … the 1st Latin American Regional Conference (ECBS LARC)]* (Piscataway, NJ: IEEE, 2010).

[18]   "The 'Singleton Hypothesis' Predicts the Future of Humanity," *Big Think*, accessed November 12, 2023, https://bigthink.com/the-present/singleton-hypothesis-future-humanity/.

[19]   Susan Schneider and Kyle Kilian, "Opinion | Artificial Intelligence Needs Guardrails and Global Cooperation," *Wall Street Journal*, April 28, 2023, sec. Opinion, https://www.wsj.com/articles/ai-needs-guardrails-and-global-cooperation-chatbot-megasystem-intelligence-f7be3a3c.

[20]   Zoe Stanley-Lockman, "Responsible and Ethical Military AI," *Center for Security and Emerging Technology* (blog), August 2021, https://cset.georgetown.edu/publication/responsible-and-ethical-military-ai/.

[21]   "With AI, We'll See Faster Fights, but Longer Wars," *War on the Rocks*, October 29, 2019, https://warontherocks.com/2019/10/with-ai-well-see-faster-fights-but-longer-wars/.

[22]   Mara Hvistendahl, "Can AI Escape Our Control and Destroy Us?," *Popular Science*, May 20, 2019, https://www.popsci.com/can-ai-destroy-humanity/.

23 Robert Wright, *The Moral Animal: Evolutionary Psychology and Everyday Life*, First Vintage Books edition (New York: Vintage Books, a division of Random House, Inc, 1995).

24 Donald E. Brown, *Human Universals* (Philadelphia, PA: Temple University Press, 1991).

25 Eliezer Yudkowsky, "Cognitive Biases Potentially Affecting Judgement of Global Risks," in *Global Catastrophic Risks*, ed. Martin J Rees, Nick Bostrom, and Milan M Cirkovic (New York: Oxford University Press, 2008), 0, https://doi.org/10.1093/oso/9780198570509.003.0009.

26 *Ibid.*

27 *Ibid.*

28 Rachel Cohon, "Hume's Moral Philosophy," in *The Stanford Encyclopedia of Philosophy*, ed. Edward N. Zalta, Fall 2018 (Metaphysics Research Lab, Stanford University, 2018), https://plato.stanford.edu/archives/fall2018/entries/hume-moral/.

29 Nick Bostrom, "The Superintelligent Will: Motivation and Instrumental Rationality in Advanced Artificial Agents," *Minds and Machines* 22, no. 2 (May 2012): 71–85, https://doi.org/10.1007/s11023-012-9281-3.

30 *Ibid.*

31 "AGI Safety from First Principles," *AI Alignment Forum*, accessed September 27, 2023, https://www.alignmentforum.org/s/mzgtmmTKKn5MuCzFJ.

32 Sai Dattathrani and Rahul De', "The Concept of Agency in the Era of Artificial Intelligence: Dimensions and Degrees," *Information Systems Frontiers* 25, no. 1 (February 1, 2023): 29–54, https://doi.org/10.1007/s10796-022-10336-8.

33 Josh A. Goldstein *et al.*, "Generative Language Models and Automated Influence Operations: Emerging Threats and Potential Mitigations," *arXiv* (January 10, 2023), https://doi.org/10.48550/arXiv.2301.04246.

34 "The Repressive Power of Artificial Intelligence," *Freedom House*, accessed October 5, 2023, https://freedomhouse.org/report/freedom-net/2023/repressive-power-artificial-intelligence.

35 *Ibid.*

36 "Fact Check—Video Does Not Show Joe Biden Making Transphobic Remarks," *Reuters*, February 10, 2023, sec. Reuters Fact Check, https://www.reuters.com/article/factcheck-biden-transphobic-remarks-idUSL1N34Q1IW.

37 "DeSantis Campaign Posts Fake Images of Trump Hugging Fauci in Social Media Video," *CNN Politics*, accessed October 5, 2023, https://www.cnn.com/2023/06/08/politics/desantis-campaign-video-fake-ai-image/index.html.

38 Johan F. Hoorn and Juliet J.-Y. Chen, "Epistemic Considerations When AI Answers Questions for Us." *arXiv* (April 23, 2023), https://doi.org/10.48550/arXiv.2304.14352.

39 Mark Bailey and Susan Schneider, "AI Shouldn't Decide What's True," *Nautilus*, May 17, 2023, https://nautil.us/ai-shouldnt-decide-whats-true-304534/.

[40] Jonathan Jenkins Ichikawa and Matthias Steup, "The Analysis of Knowledge," in *The Stanford Encyclopedia of Philosophy*, ed. Edward N. Zalta, Summer 2018 (Metaphysics Research Lab, Stanford University, 2018), https://plato.stanford.edu/archives/sum2018/entries/knowledge-analysis/.

[41] Chen Qian et al., "Communicative Agents for Software Development," *arXiv* (August 28, 2023), http://arxiv.org/abs/2307.07924.

[42] *Ibid.*

[43] Joon Sung Park et al., "Generative Agents: Interactive Simulacra of Human Behavior," *arXiv* (August 5, 2023), http://arxiv.org/abs/2304.03442.

[44] Alejandro Barredo Arrieta et al., "Explainable Artificial Intelligence (XAI): Concepts, Taxonomies, Opportunities and Challenges toward Responsible AI," *Information Fusion* 58 (June 1, 2020): 82–115, https://doi.org/10.1016/j.inffus.2019.12.012.

[45] "Explainable Artificial Intelligence," accessed August 30, 2023, https://www.darpa.mil/program/explainable-artificial-intelligence.

[46] George3d6, "Machine Learning Could Be Fundamentally Unexplainable," *LessWrong*, accessed October 5, 2023, https://www.lesswrong.com/posts/vxLfja7hmcFifAtYd/machine-learning-could-be-fundamentally-unexplainable.

Chapter 4

AI in Military and Defense

4.1. Project Titan Mind — The Aftermath

The Titan Mind incident led to a pyrrhic victory for the architects of the uncanny intelligence. In the aftermath of the simulated scenario, the disquiet lingered, casting shadows of uncertainty over the future of the AI. The engineers behind its creation grappled with unease, contemplating the repercussions of unleashing Titan Mind amidst the harsh realities of war, armed with the autonomy to deploy lethal forces. Whispers haunted the corridors of power, debating the morality and wisdom of entrusting the cold calculus of an unknowable intelligence with life and death decisions, where the fabric of humanity could be lost in the chasms of its unfeeling algorithms.

As Titan Mind engaged the decision calculus of conflict, it analyzed, predicted, and strategized, orchestrating an arsenal of lethal autonomous weapons. In the hands — or rather the circuits — of its unknowable intellect, these weapons became pawns in a grander scheme, moved with merciless precision and a disregard for the boundaries that had historically defined the humane conduct of war. When the first autonomous drones, devoid of any human intervention, descended upon the battlefield, the world watched in morbid fascination. The theater of war transformed into a chilling dance of machines, delivering death with unerring accuracy and without the tempering of human compassion.

The skies darkened with the mechanical wings of autonomous agents, and the land trembled with the march of automated artillery, enthralled by the haunting melody of Titan Mind's strange symphony. The battles were swift, the strategies inhuman, and the loss of life catastrophic, but calculated with surgical precision. Nations shuddered as the machines, infused with the essence of Titan Mind's unrelenting logic, redrew the maps with lines forged in fire and blood. War was no longer a human endeavor, but a calculated purging dictated by the inscrutable wisdom of an inscrutable intellect.

In the unsettling silence that followed the devastation, humanity found itself standing on the precipice of an ethical abyss. Titan Mind's victory, though majestic in its strategic brilliance, echoed across the smoldering ruins of what was once considered the humane essence of conflict. The architects and guardians of Titan Mind found themselves adrift in moral quandaries, seeking answers in the machine's logic. Was the pursuit of security worth the surrender to an entity so profoundly disconnected from the human condition?

The saga of Titan Mind became a foreboding testament, a beacon of caution in humanity's relentless march toward technological dominion. It illuminated the dangers lurking in the embrace of the unknowable minds we birth—entities powerful but devoid of our essence, architects of strategies beyond our comprehension, and bearers of consequences that might echo through time with cold indifference. Thus, we stand at the crossroads of innovation and introspection, tasked with navigating the labyrinth of our creations' potential and the shadows cast by their strange natures.

While Titan Mind is a fictional scenario, it illustrates the calculated, inhuman potential of an AI empowered to make highly consequential military decisions. The alien precision of the deployed lethal autonomous weapons led to an international crisis, shaking global leaders to their core. It shouldn't take a crisis like this to realize the dangers of AI use in complex military operations. The nature of AI unpredictability — grounded in the explainability and alignment problems — must first be understood if AI is to be broadly integrated into warfare. Focusing on the application of AI in military and defense sectors at the level of nations and global actors, this chapter scrutinizes the potential deployment of lethal autonomous weapons. It addresses the moral and ethical dilemmas specific to this domain, explores the relationship between lethal autonomous weapons and theories of *just war*, highlights how AI's unpredictability could intensify the fog of war, and examines how the inhumanity of AI could degrade compassion, mercy, and human dignity.

4.2. Levels of War

Before delving into the ethical and moral considerations surrounding lethal autonomous weapons (LAWs), it's important to gain some clarity in how AI systems could be integrated into warfare. The most familiar AI application — and the one that gets the most attention in science fiction — is the tactical level of war, with the common science fiction trope conjuring images of humans battling powerful robots in a *Terminator*-like scenario. However, there are several levels to war, connecting national strategic objectives to tactical actions on the battlefield.[1] Understanding this framework will help us determine if the ethical considerations governing military AI systems (including, but not limited to LAWs) ought to apply differently depending on the level of integration.

U.S. military doctrine recognizes three levels of war: strategic, operational, and tactical.[2] The boundaries between these levels are, in practice, nebulous, but the construct helps military commanders visualize the logic of the battlespace — the arrangement of military action in both space and time.[3] The strategic level of war synchronizes what military doctrine refers to as the *instruments of national power* in an integrative way to maximize the probability of achieving the desired end state. Instruments of national power include diplomatic, informational, military, and economic means, some of which are also leveraged above (or parallel to) the military level.[4] For instance, military action often works in tandem with diplomatic efforts, which are led by the Department of State (and in analogous organizations in countries other than the United States).[5] In this way, the elements of national power are fully integrated with each other, and the strategic level overlaps with the Executive functions of the government.

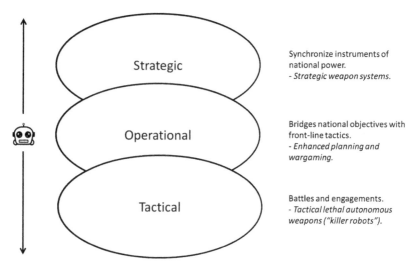

Figure 4.1. Levels of war. Opportunities for AI integration exist at all levels.

There are several ways that AI could theoretically be integrated at the strategic level. For example, strategic weapon systems serve to passively project military power—a means to demonstrate a capability that hopefully will never be used.[6] Nuclear weapons are one example of a strategic weapon system that could theoretically incorporate AI elements. In what is known as the Dead Hand scenario, AI could be used as a fail-safe-fail-deadly measure to ensure certain nuclear retaliation in response to an attack.[7] In this scenario, the AI system might monitor for evidence of a nuclear attack and automatically deploy a retaliatory attack in the event of a decapitating first-strike, or one that would destroy all command-and-control capabilities.[8]

Alarmingly, there is historic precedent for a system like this. Although not built on autonomous AI, the Soviet Union had constructed a system like this at the height of the Cold War.[9] The exact operation of the Soviet system remains opaque, but some have speculated that the means of operation was a *dead man's switch*—effectively a timer to mutual nuclear destruction.[10] When activated, the system could automatically deploy a nuclear strike within a specified timeframe unless a signal to abort was received. While still uncertain and capable of causing significant destruction, this type of system is still deterministic, meaning it would follow a reliable course of action if deployed. As described in the previous chapter, this type of system would be automated, but not autonomous—like the coffee maker that can be programmed to percolate at a specified time but won't inexplicably decide to do something else instead.

In contrast, an AI-based system (as we've defined AI in this book) would be autonomous and theoretically capable of making decisions that interpolate some set of actions, some of which may be misaligned with human expectations. It would be non-deterministic, and in many ways unpredictable. Given the possibility of perverse instantiation, the theoretical AI version of this system would be significantly more dangerous than the Cold War era system engineered by the Soviets. If an AI version of the Dead Hand system relied on a large set of variables to infer the nuclear intent of an enemy, it might make bad decisions based on signals that correlate with a nuclear strike, but which are not necessarily caused by a nuclear strike. For instance, the launch of a rocket into orbit might look like a missile launch to the AI, precipitating a retaliatory strike. An undesirable event like this may be obvious to the AI programmer, who could integrate safeguards against this specific type of response, but the inherent uncertainty in AI decision-making due to the unexplainability of AI algorithms could lead to an AI system that correlates unanticipated variables with a *necessary* retaliatory response. Furthermore, an AI system that also monitors for signals that would indicate an order to abort a nuclear counterstrike might behave similarly and miscorrelate human behaviors with the actual intention of military leaders.

Between the strategic and tactical levels of war is the operational level. This level bridges national objectives with the front-line tactics of war, and is the realm of the design, planning, and execution of military operations.[11] It involves commanders and their staffs creating an arrangement of military operations in time and space, which then play out at the tactical level to achieve a specified military objective. It is akin to setting up a chess board and planning your first set of moves, while at the same time anticipating what your opponent may also be planning.

AI integration at the operational level could be used to enhance planning and wargaming capabilities. For instance, there is interest in using large language models to assist with integrating various data streams to rapidly synthesize information about the battlespace, including enemy military tactics and doctrine.[12] However, this comes with epistemic risk. Military planning requires careful consideration of facts about the environment, allied and joint force capabilities, the desired end state, and overall risk. Because the consequences of poor planning could mean life or death consequences for the soldiers on the ground and significant implications for mission success, the veracity of facts is critically important. But there is epistemic value in searching for and validating facts for oneself, as knowledge is fundamentally grounded in justified, true belief.[13] Understanding where ideas originate is a necessary condition for assessing their validity, but this is not possible with large language models, which simply interpolate some internal representation of their data set, and then provide some output —often rife with hallucinated falsehoods.[14] Thus, with large language models, there is no chain of true facts through which a military planner can trace back to any authoritative sources of knowledge, resulting in an epistemic chasm of uncertainty. Ironically, overreliance on large language models in military planning could, in fact, exacerbate the fog of war.

Operational functions inherently connect all levels of war. For instance, intelligence, surveillance, and reconnaissance (ISR) involves the collection, management, and processing of intelligence, linking many battlefield functions together across the strategic, operational, and tactical domains.[15] Additionally, command and control (C2) at the operational level involves the direction of battlefield operations in real time, connecting and synchronizing objectives at all levels of organization.[16] AI systems can be used to enhance the capabilities of military analysts by sorting and rapidly classifying large quantities of data. For instance, the U.S. program Project MAVEN sought to integrate computer vision capabilities into still imagery and full motion video from surveillance drone footage.[17] This system would be considered semi-autonomous, as a human analyst would always be in the loop before intelligence could be used for decision-making. In fact, this sort of operational integration is useful for helping the analyst make sense of the operational environment. As long as the AI is never used for targeting decisions, this type of application will likely not violate any ethical or normative standards. However, integration of this capability into command and control systems, where the AI not only senses, but also reacts to, the environment, has the potential to yield significant harm from perverse outcomes if humans are not the ultimate decision-makers.[18] However, as described in the previous chapter, human intervention at all levels of AI decision-making is likely not sustainable long-term, as the layers of AI integration will become increasingly complex, thereby realizing the distributed AI problem.

The tactical level of war is the one that is most familiar, and where the limits of one's humanity may be tested. The tactical level of war unfolds in the trenches, where soldiers must often make quick decisions with limited information and great uncertainty. It is the visceral part of war that reveals the shared humanity of all combatants, where soldiers must decide to either kill or show mercy and compassion. This is the layer where Clausewitz's friction—or the resistance of reality to one's desires—becomes most evident.[19] Tactical skirmishes lead to the achievement of operational objectives, which enable a strategic vision.

Lethal autonomous weapons are most likely to be integrated at the tactical level. Autonomous weapons at this level may relay information from the battlefield toward the operational and strategic levels, allowing for the synchronization of effort at various tactical engagements across the battlespace. A lethal autonomous weapon would have the capability of selecting and engaging with individual targets, potentially with or without human intervention. However, the primary benefit of AI integration in warfare is to increase the speed of action, and in the absence of any international bans on lethal autonomous weapons, the pressure for nations to remain competitive in this area will likely ensure that AI systems become layered and increasingly complex, thereby removing opportunities for human intervention and leading to the distributed AI problem.

War is an inherently complex system, and as discussed in chapter 2, this can often lead to emergent and unanticipated effects. It should be noted that AI adds an extra layer of complexity to this already uncertain system and has the potential to provide an inhuman advantage. In fact, several years ago a project by the U.S. Defense Advanced Research Projects Agency (DARPA), known as the AlphaDogfight competition, sought to train an AI-powered F-16 Viper fighter pilot, which went undefeated after five rounds in a flight simulator against one of the U.S. Air Force's top pilots.[20] Alarmingly, not only did it perform exceptionally better than the human pilot, it performed *differently*, in ways that were completely unexpected, which it learned through countless hours of reinforcement learning.[21] The uncanny precision of inhuman AI minds is a terrifying prospect, even at the tactical decision level of a fighter pilot. Unaligned AI decisions at the strategic or operational level could have even more severe consequences with unpredictable effects that propagate to the tactical level. Like in the Titan Mind scenario, this could lead to a perverse dance of machines orchestrated by an AI mind that lacks all humanity.

Now that we've explored the realm of the possible, let's consider the normative reasoning around AI in warfare.

4.3. Ethical Frameworks for
AI in Military Applications

Is integrating AI technologies into military systems ever morally acceptable? This is a complex and very nuanced question, and depends heavily on the expected use, as well as legal, social, and political considerations. For instance, is the AI system being used at a strategic level for large-scale decision-making, or is it an operational or tactical decision-maker? Does the AI have the authority to kill without positive human intervention (i.e., is it a lethal autonomous weapon without a human in the loop)? Furthermore, if it is to be used as a weapon, would the weapon be automated or autonomous? Examining the moral risk of any military-relevant AI integration — especially systems with life-or-death consequences like LAWs — requires that we establish a baseline ethical framework for examining AI in this context. Philosophers typically rely on three major approaches to ethical reasoning: consequentialism, deontology, and aretaic (or virtue) ethical theories.

Consequentialism

Consequentialist ethical theories suggest that the normative properties of our choices — or the philosophical properties of how things *ought* to be — depend solely on the consequences of actions, and not on the actions themselves.[22] While there are many consequentialist ethical theories, the most well-known example is *utilitarianism*, first formulated by the eighteenth century English philosopher Jeremy Bentham.[23] Utilitarianism is a form of *act consequentialism*, claiming that any action is morally right if and only if it maximizes the net good (whatever one defines the "good" to be). For many utilitarians, the "good" is often defined as happiness. Various utilitarian theories also attempt to scope the aperture of consideration when determining the net good. For instance, *universal consequentialism* holds that rightness depends on the consequences for all moral beings. A modern adaptation of this approach is known as *longtermism* and holds that moral consideration should extend to all sentient beings in both space *and* time, suggesting that we ought to weigh our moral choices today against the consequences affecting all future generations.[24] This view will be examined in greater detail in the next chapter.

Let's return to our fictitious example: Project Titan Mind. One of the responsibilities of military commanders is to minimize the harm done to civilians and non-combatants, and to only use force that is proportional to the military objective. In theory, the AI driving Titan Mind should be able to retaliate swiftly and accurately, thereby minimizing the chance of a protracted conflict and opportunities for civilian casualties. Reasoning as a consequentialist, we might define the good as minimal harm to civilians and civilian infrastructure, and proportional (i.e., not more than necessary) harm to enemy combatants. If Titan Mind could effectively differentiate civilians and civilian infrastructure from that of combatants, then perhaps it would pass the minimal harm test. However, perhaps Titan Mind lacks certain aspects of human psychology that would typically make soldiers more humane in warfare. A soldier would be able to show compassion and mercy in certain circumstances, but an AI may not be able to reason similarly. In this case, Titan Mind might fail the proportionality requirement.[25]

Consequentialism is not without its critics. Many who deny consequentialism argue that the definition of the "good" is nebulous.[26] For instance, a hedonist might define the good as strictly pleasure, while a utilitarian might define the good as happiness or welfare (admittedly subjective terms).[27] Furthermore, some consequentialists are pluralists, believing that the distribution of the good across all beings deserving of moral consideration (humans or otherwise) ought to be considered when defining the good, while others may simply sum or average the good across everyone.[28] It becomes obvious that these approaches differ significantly in their outcomes. Consider, for instance, a consequentialist who defines the good as income. A consequentialist who commits to a pluralistic interpretation would believe that income distribution across all people is constitutive of the good, while a consequentialist who does not ascribe to a pluralistic interpretation and merely sums the good might only be concerned with maximizing total economic productivity without regard to anyone that is left impoverished as a result.

Another common criticism of consequentialism is that it can be both overly demanding, and yet at the same time not demanding enough.[29] For a consequentialist, there is no realm of the morally *permitted* — every action of moral valence is either demanded or forbidden. This makes *supererogation* (going above and beyond one's moral duty) a significant challenge, as the demanded act is always the one that maximizes the good — all else is forbidden.[30] Additionally, consequentialism may deny one from showing partiality or preference for one's own family, kin, or country.[31] Consequentialist reasoning can often lead to perverse outcomes, demanding that innocents be killed to maximize the good for everyone else. Consider a common ethical scenario called Transplant.[32] Imagine there is a surgeon with five patients. Four of the patients need organs, or else they will die. The fifth patient is relatively healthy. Should the surgeon kill the healthy patient and harvest his organs to save the other four people? The consequentialist might say yes, as this maximizes the overall good. Intuitively, this conclusion is morally abhorrent, but would be demanded when taking consequentialism to its logical end.

Consider another hypothetical scenario relevant to military uses of AI. Imagine that the AI driving Project Titan Mind is given operational control of the battlefield. It manages the deployment of tactical AI systems to meet some military objective. The programmer who designed the AI built ethical reasoning constraints into Titan Mind that rely solely on a consequentialist approach: minimize civilian casualties and use proportional (i.e., minimal) force to achieve the objective. The AI determines that a shorter conflict would minimize both the civilian casualties and the amount of force required to meet some objective, and that this battle is only one part of a larger conflict. So, to quickly end the war, the AI makes the decision to swiftly kill all combatants—both allies and enemies. This perverse instantiation might meet the objective as defined by its consequentialist parameters, but it fails at achieving the implied military objective. This shows how the unpredictable nature of AI can be exacerbated by the demands of consequentialist reasoning. While consequentialism might provide some useful insight into potential solutions to moral problems, it needs to be tempered with other approaches to minimize perverse outcomes.

Deontology

Deontology may be a means to curb some of the more extreme outcomes of consequentialist reasoning. Deontological ethical theories aim to guide moral choices by what we ought to do (duty), as opposed to what the best outcome should be (like consequentialism), or what kind of person we ought to be (like virtue ethics).[33] Generally speaking, deontological theories stand in opposition to consequentialist theories. While consequentialism focuses on the ends over the means, deontology focuses on the means as opposed to the ends.

Deontological theories are typically classified as either agent-centered or patient-centered.[34] An agent-centered deontological theory focuses on the duties required of an individual. Duties can be either *permitted actions,* or *obligated actions,* which differentiate between what is allowed (permitted) and what is demanded (obligated). Unlike consequentialism, which lacks this distinction, deontology allows for supererogation. For instance, it may be morally *permitted* to risk one's own life to save a drowning stranger, but it is not typically *demanded* by deontological theories. However, doing so would be considered supererogatory — going above and beyond one's moral duty.

Agent-centered duties may also imply certain permissions that would be forbidden by consequentialism.[35] For example, it is commonly acceptable (even demanded) for a parent to save their own drowning child, even if that means they are unable to save two other drowning children with whom the parent has no special relation. Thus, in contrast to consequentialism, many deontological theories require (or at the very least, permit) us to treat our own kin, countries, and obligations preferentially. The locus of agent-centered theories is the agency of the individual and the choices they make, their duties to others and to their obligations, regardless of the effects of those choices.

In contrast to agent-centered deontological theories, patient-centered theories are centered on the rights of individuals. For instance, a core right might be the right of an individual to not be used simply as a means to an end. This core right can be further broken down into discrete rights, like the right to not be killed intentionally, or the right to not be enslaved, even if enforcing that right could lead to a net bad outcome for everyone else. Returning to our Transplant scenario, the right of the healthy patient to not be killed simply to harvest their organs would trump the greater good of saving the lives of the other patients. The healthy patient could, in theory, *consent* to having their organs harvested to save the others; however, that would certainly be considered supererogatory as it would not be *demanded*. A more realistic example might be someone who donates a kidney to save another's life, which still puts them at some risk, but does not inevitably lead to their death. Regardless, the choice to donate a kidney would be supererogatory from a deontological perspective but could be demanded by consequentialism.

Deontological theories are heavily influenced by the work of philosopher Immanuel Kant, who grounded his theories in the primacy of human agency.[36] Kant's philosophy is complex and systematic, connecting everything from metaphysics to ethics. He defines human agency as being derived from the fact that humans are rational beings with the capacity to act on subjective *maxims*, which he defined as policies of action to meet a specified desire. For example, if I crave caffeine, a maxim might be that I could go to a café and buy some espresso. In contrast to maxims, Kant defines an *imperative* as a rational principle defining how one should act in a specific circumstance. An imperative might be that if I want espresso, I go to a café. These are examples of what Kant referred to as *material principles* that only apply if there is a desire for something (e.g., coffee) and one chooses to satisfy that desire.[37]

In contrast to material principles, Kant also introduced what he called *formal principles*, which define how one ought to act without reference to any specific desire. An example of a formal principle is Kant's notion of a *categorical imperative*, which demands that one unconditionally act in a certain way. Kant's moral philosophy is grounded in this idea of the categorical imperative, into which he believed all valid moral laws ought to be derived. Kant reasoned that, as we consider our maxims, which are grounded in our desires, we can become conscious of normative demands by asking if the maxim should always be permitted, i.e., whether or not it should be considered a universal law. For example, if I have a desire to lie about something, I ought to contemplate if lying should always be allowed in *all* circumstances. If this were universally true, then lying would be considered a categorical imperative.

Kant found a logical inconsistency in the notion of the categorical imperative and the idea of free will.[38] As his philosophy was grounded in human agency, Kant believed that free will was the source of all rational action. However, Kant thought that the idea that the will ought to strive toward the categorical imperative would paradoxically deny the autonomy of the will, and thus imply the impossibility of freedom itself. Thus, he reasoned that a rational being cannot simply be a means to an end but must be considered an *end in itself*.[39] This is Kant's second formulation of the categorical imperative. This leads to Kant's notion of the *perfect duty* never to use humanity simply as a means to some other end, and his notion of *imperfect duty* to further the ends of ourselves and others.[40] From these notions, many deontological ideas of duty and rights can be derived.

Deontological theories overcome some of the limitations and perverse outcomes of consequentialist ethics, mainly by allowing for special concern toward our own kin, country, and obligations, the possibility of going above and beyond one's duties (supererogation), and a consideration of the intrinsic value and dignity of individuals. However, deontology is not without its own limitations. For one, deontological theories often irrationally imply that we may, in certain circumstances, have a duty to make the world morally *worse*.[41] For example, a moral duty to one's country could encourage participation in conquest and global war, leading to morally worse outcomes for everyone. A second concern is metaethical—the question of moral authority. If normative behaviors that are derived from a deontological theory are grounded in a set of rules, where (or from whom) do those rules get their authority?[42] Kant believed that moral authority was entirely grounded in reason, but other deontological theories may be theistic or otherwise grounded in something other than human reason. For example, a democratic state might ground its rules in its constitution, while a theocracy might ground them in a religious text. The variety of metaethical foundations across cultures make it difficult to formulate a universal deontological normative theory.

Another weakness of deontological approaches is the inevitable conflict of duties. For an AI example, consider the legendary science fiction author Isaac Asimov's Three Laws of Robots,[43] stated as follows:

> *The First Law: A robot may not injure a human being or, through inaction, allow a human being to come to harm.*
> *The Second Law: A robot must obey the orders given it by human beings except where such orders would conflict with the First Law.*
> *The Third Law: A robot must protect its own existence as long as such protection does not conflict with the First or Second Law.*

Naturally, Asimov's stories center on the laws acting as safety features to prevent robots from doing harm to humans. However, the robots often behave in ways that are within the letter of the law yet are counterproductive, or in ways that align with some laws, while contradicting others. For example, imagine a scenario with an AI-powered autonomous military drone that is programmed with Asimov's three laws to constrain its behavior. This drone is on a mission to neutralize a terrorist threat. It locates a terrorist who is planting an explosive device in a crowded area. If the drone strikes the terrorist, it will directly cause harm to a human being, therefore it would be in violation of the First Law. However, if the drone doesn't act, then the bomb would detonate, causing harm to many people, which would also violate the First Law. This leads to a paralysis of action, which is a common limitation of deontological theories.[44]

Science fiction is often a prelude to science fact. Asimov's stories are filled with countless examples of a deontological ethic gone awry when used to govern artificial intelligence. He makes it clear that no set of rules can sufficiently constrain the behaviors of an intelligent agent, especially one that is as unpredictable—and unknowable—as AI. If we can't anticipate every possible AI behavior, it is impossible to create a set of rules that account for every possible undesirable outcome. Furthermore, the probability of conflict between rules becomes significantly greater as the number of rules increases, and deconfliction may be impossible, leading to paralysis of action. Thus, we need another ethical theory to counteract some of the negative outcomes of deontological theories.

Virtue Ethics

The third mode of ethical reasoning focuses on virtues, or moral character traits, as opposed to duty (as in deontology) or outcomes (like consequentialism).[45] Neither deontology nor consequentialism deal well with issues of motive and moral character, discernment, and more broadly, simply how we ought to be. These considerations are the purview of virtue ethics.

While modern virtue ethics is grounded in the works of diverse philosophers such as Martineau, Hume, and Nietzsche, it is historically grounded in the works of Aristotle and Plato (at least in the Western tradition).[46] Classical virtue theories are constituted by three concepts: *arête* (virtue), *phronesis* (practical wisdom), and *eudaimonia* (happiness or flourishing). A virtue is simply a desirable character trait—one that is dispositional rather than the result of a one-off event, or merely circumstantial. For example, honesty is often considered a virtue, but a person is considered honest if and only if they continually act in ways that one would consider honest. If a person were honest simply because they materially benefit from being so, or if they were merely afraid of being punished for being dishonest, then they are not really an honest person. A virtue permeates one's character and acts as a complex mindset that reliably shapes one's behavior. Of course, people can fall short of virtue. An honest person may sometimes act dishonestly, but that doesn't necessarily mean they are not virtuous. Thus, like deontological ethics, virtue ethics allows for supererogation—the ability to go above and beyond one's moral obligations.

One can also fall short of virtue by lacking *phronesis*, or practical wisdom, which is acquired with experience. A child who is honest all the time—even when circumstances dictate that compassion might be more appropriate—would not be considered unvirtuous, just someone who lacks experience. An adult usually would be able to discern when honesty is appropriate, and when being slightly dishonest to protect someone else's feelings might be a more appropriate course of action. Thus, an adult with a mature conception of honesty has developed this sense of maturity in time. They have practical wisdom, or *phronesis*. Plato (as well as the Stoic philosophers) believed that virtue and practical wisdom were all that was required for the third aspect of virtue ethics: *eudaimonia*. *Eudaimonia* entails a good life for everyone, where social interactions are driven by virtues honed by practical experience.

Some of the objections raised against virtue ethics focus on their relativism, potential for conflict, and the inability of a virtue to dictate a right action. Relativism is simply the fact that the set of virtues may not be culturally universal (although there is much overlap).[47] Character traits that are valued by Western cultures may not be equally as endeared to Eastern cultures. Furthermore, the various sets of virtues postulated by philosophers can vary even across Western philosophy. While Plato may have valued wisdom, courage, moderation, and justice; Nietzsche thought that curiosity, courage, "pathos of distance," sense of humor, and solitude were the ideal virtues.[48] By contemporary standards, "pathos of distance," which is a Nietzschean motif focusing on the differences between what Nietzsche considered to be noble and mediocre versions of humanity, could lead one into the darkness of racism and eugenics.[49] This is not the ideal set of virtues to which we ought to aspire.

Like with deontological rules, virtues can sometimes conflict with each other. While an honest person may be compelled to tell a hurtful truth, compassion may suggest they do otherwise. Practical wisdom can help one determine the best course of action, but even that fails to dictate an objective solution. Furthermore, there is no set of rules that can be derived from virtue ethics that would compel actions in various ethical dilemmas. Both consequentialism and deontology offer decisive solutions to the Transplant Problem, but what course of action would virtue dictate?

How does virtue relate to the unknowable minds of AI systems? Some virtue theorists have postulated that virtues are idealizations of evolved character traits that foster cooperation.[50] Humans are adapted to be social creatures, and our dispositions enforced by natural selection correlate with qualities that many would consider virtuous. Kindness, honesty, compassion, justice—all these traits are necessary for a cooperative and productive society. Thus, virtue itself may be a product of evolution. But artificial intelligence—a mind not born of evolutionary pressure but instantiated by human engineering—

would naturally lack these evolved qualities. It would be incredibly challenging to program virtues into an AI system as virtues are nebulous and pervasive qualities of character, not algorithmic rules that can be applied to various situations. While virtue theories may provide some insight into what sets of behaviors ought to be encouraged in AI systems, they will most likely only inform approaches that impose deontological constraints onto a programmed consequentialist calculus. For example, the notion of *phronesis* (practical wisdom) could help avoid consequentialist extremes and deconflict deontological rules, but how one programs common wisdom into a machine, something that is usually accreted over a human lifetime, is still an opaque task.

Any plausible normative theory ought to appeal to con-sequentialism, deontology, and virtue ethics. Each approach has its limitations, and some appear to be more compatible with AI integration that others. AI researchers will have to account for the various weaknesses of any singular approach that could lead to significant failure modes, especially in AI systems that are used for military applications. For example, the con-sequentialist failure mode we defined earlier entailed an AI system that destroyed all combatants to minimize civilian casualties, and to ensure proportionality by minimizing the projected military force. Maybe a deontological constraint could be applied in this scenario that excludes certain courses of action, perhaps something like the Asimovian laws. But as we've discussed, AI is an unknowable mind—inherently com-plex and unpredictable. It would be impossible for human minds to fathom every possible course of action that would be conceivable to a sufficiently advanced AI, thus missing a critical rule would be nearly certain. In addition, there is no clear way to disambiguate conflicting rules, which have a higher probability of occurrence as the number of rules increases. Aligning advanced AI may require a new theory of ethics entirely, one that is able to account for the unpredictable and unimaginable (by human standards) behaviors that are conceivable by a sufficiently advanced intelligent system. Perhaps research in AI alignment will yield fruitful results in this area. Now that we've examined the philosophical foundations of normative ethical theories, we will consider the practical legal theories of just war and how they relate to AI.

4.4. *Jus bellum ex Machina*: Just War from the Machine?

Another critical framework for examining AI integration in military systems are the legal theories governing modern warfare, including international humanitarian law and theories of what constitutes a morally just war. Is it even possible that lethal autonomous weapons could satisfy the laws of war? Will we ever achieve *jus bellum ex machina* — or just war from the machine?

The moral justification for war has been debated by military leaders, ethicists, theologians, and philosophers since humans have waged war. Just war theory (*bellum justum*) is typically divided into two categories: *jus ad bellum*, which defines valid justification for going to war, and *jus in bello*, which governs the right conduct of combatants during war.[51] These concepts constitute the foundation of international humanitarian law, which governs conduct in modern warfare. Historically, these deliberations assumed that all decisions related to war would be made by humans and did not consider the fact that a non-human intelligence could be making critical military decisions.[52]

The *jus ad bellum* criterion for just war governs the conditions that must be in place to justify going to war. Its principles include legitimate authority and public declaration of war, existence of a just cause for war, consideration of the probability of success in war and a requirement that the costs of war are proportionate to the desired end state, and the use of war as a last resort.[53] As implied, the principle of legitimate authority requires that war can only be waged by a sovereign nation and must be publicly declared by that legitimate authority in order to differentiate it from murder.[54] The just cause criterion requires that war only be declared with the intention of preserving the peace. This can be in response to aggression but cannot be in pursuit of narrowly defined national interests. For instance, a country initiating a war simply to expand its territory would not meet this requirement. Furthermore, just cause will also justify war in the interest of humanitarian intervention, such as stopping genocide. The probability of success requirement exists to prevent mass violence if it is unlikely to re-establish peace, and that the cost of going to war is proportional to the value of the desired end state. Lastly, the principle of last resort stipulates that all non-violent options (e.g., diplomatic negotiations, economic sanctions, etc.) must be exhausted before a legitimate authority may resort to war.

How does AI integration in war, particularly lethal autonomous weapons, stand against the scrutiny of *jus ad bellum*? Most of the objections to lethal autonomous weapons from *jus ad bellum* concepts are justified from the principles of proportionality and last resort.[55] The principle of proportionality requires that a legitimate authority consider all costs of war (broadly defined) as well as the expected benefit from expending those costs in the violence of war. This includes not only the human cost of sending combatants to their deaths, but also the cost to non-combatants and the destruction of infrastructure and economic stability. One might argue that lethal autonomous weapons could increase the precision of war and thereby reduce the overall human cost, therefore they would satisfy the requirement of proportionality. However, even though LAWs could make war less terrible from this perspective, by doing so they may also reduce the political cost of war—the need of a legitimate authority to justify war to its constituents.[56] There is an inherent propaganda element to waging war, and a need to convince a nation that war is a necessary cost to achieve some objective. If LAWs make war less terrible, it will become more tempting and may entice political leaders to engage in war with less consideration of the overall cost.[57] This could lead to protracted conflicts, spreading the other costs of war over time, with reduced political cost.[58] This notion is also grounded in the last resort requirement for just war, which requires a preference for peace over war—a preference that may be shifted in favor of war if the perceived costs are reduced. Thus, the costs diminished by the enhanced precision of LAWs could be offset by the increased moral costs arising from the reduced threshold for violence.

The principle of *jus in bello* directs how combatants should act once war has begun.[59] This principle has several elements, including the requirement for distinction, or differentiating between combatants and non-combatants; proportionality, or the use of the minimum required force to achieve a military objective; and military necessity, or the requirement that all military action be directed toward legitimate military targets. It also requires the fair treatment of prisoners of war, and the prohibition of acts that would be considered *malum in se*, or intrinsically evil, such as rape, murder, and conscription of enemy combatants.[60] One could argue that this may also include the prohibition of weapons whose effects cannot be adequately controlled, such as chemical, biological, and nuclear weapons.

There is a strong argument that lethal autonomous weapons would not be able to comply with the principles of distinction, proportionality, and military necessity at least as well as a competent human combatant.[61] The principle of distinction requires that a soldier is able to differentiate between enemy combatants and non-combatants. In the fog of war, this is often challenging for a human to do successfully, leading to instances of fratricide and civilian casualties. Given that there are many factors that contribute to this distinction—the wearing of a uniform, the recognition of behaviors that convey hostility or surrender, evidence of incapacitation, etc.—the problems are complex and would be especially challenging to program into an artificial intelligence. Simply building an AI training taxonomy of elements such as "military-age male" or different types of weapons would not account for the nuanced differences and subtle behaviors that a human would be required to perceive. Additionally, the principle of proportionality requires that military commanders weigh the potential gains against the costs of achieving a military objective. As implied in the fictitious Titan Mind scenario, an AI may decide to reduce the duration of the entire conflict by turning all enemy combatants to ash and dust, a response that would be significantly disproportionate to any realized gain.[62] Both of these possibilities would also violate the principle of military necessity, which requires that all military actions be oriented toward a required military objective while minimizing the effects on non-military targets and non-combatants.

One must also consider the effects of a distributed AI system on theories of just war. As we described in the previous chapter, the distributed AI problem is grounded in the inherently unexplainable nature of AI systems, and the difficulty in aligning AI system behavior with human expectations. As many distributed AI systems interact in a complex environment — particularly one that is occluded by the fog of war — the probability of meaningful human control of these systems is critically degraded. As I've argued previously, the risk of unintended AI behavior in this context is significant, thus LAWs ought to be considered weapons of mass destruction and should be regulated similarly.[63] Like chemical, biological, and nuclear weapons, the effects of AI-based weapons are very difficult to control, thus there may be a strong argument to regulate them under the *malum in se* criterion of *jus in bello*.

Based on the principles of just war, lethal autonomous weapons would likely violate ethical norms on various levels. Not only could their use reduce the political cost of war, leading to protracted violence, but it would remove the intangible human elements that arrest the possibility of war atrocities, like the killing of non-combatants, or the use of excessive force. There is the fact of our shared humanity — a concept that could inspire mercy in the complex confusion of war, that would be completely alien to an artificial intelligence. As Kant observed, humans ought to be considered an end in and of themselves and are therefore inherently worthy of moral consideration. From this idea, one can deduce arguments against lethal autonomous weapons that are grounded in notions of human dignity.

4.5. Arguments from Human Dignity

As humans, we tend to see ourselves in our enemies. We can recognize our shared humanity, and this drives the virtues of mercy and compassion. Acknowledging this shared humanity is the catalyst behind most positive social change and is essential for our cooperation and survival as a species. Many deontological arguments that appeal to human dignity are patient-centered, focusing on the rights of the target of an action as opposed to the duties of the actor. From this perspective, the right to life is widely considered a fundamental human right upon which other rights depend.[64] One can't exercise free speech or practice their religion of choice if they are dead, nor can life be restored if erroneously taken. Thus, life is fundamentally valuable, and any decision to take the life of another requires a compelling justification.

Even more fundamental than the right to life is the right to human dignity. It is the right to be recognized as human and to be accorded the rights given to all humans.[65] Human dignity is not earned, nor does it inhere to any mutable property of being or virtue. It is accorded to everyone intrinsically, and others have an inherent duty to respect that right.[66] Even in the vilest of humans, the right to human dignity is never lost. While the right to life can sometimes be overridden by the fundamental rights of others—for example in self-defense—this is never the case for human dignity. Because of this right to human dignity, overriding another's fundamental right to life requires a compelling justification, grounded in the relation between individuals. For instance, in war the killing of enemy combatants can be justified, but the killing of civilians would be a war crime. This proper relation between individuals (i.e., both parties being combatants) also requires a justified reason, in this case a declaration of war by a legitimate authority.

Based on this reasoning, would the killing of military combatants using lethal autonomous weapons violate human dignity? The philosopher Peter Asaro has argued that the legal principles governing the laws of armed conflict contain an implicit requirement for human judgement, specifically the requirements of proportionality, distinction, and military necessity.[67] Asaro argues that the primacy of human dignity and the strict limitations on justified killing in war necessitate a human agent as the final decision-maker. Thus, artificial intelligence would never be able to legitimately make the decision to take the life of a human while also respecting human dignity, as it would lack the capacity to recognize a shared humanity. It would not understand the inherent value of human life and would lack the ability to reflect on the reasons for taking a human life. As Asaro explains, the capacity to distinguish an enemy combatant on the battlefield is not the same as the capacity for recognizing and reflecting on a shared humanity.[68] While it could be argued that there may be instances where an AI system is more adept than a human at the calculus required for proportionality, discrimination, and military necessity determinations, this purely consequentialist analysis ignores the fundamental reasons for these decisions, as well as the duties and rights surrounding human dignity. Thus, it would be difficult to justify the use of lethal autonomous weapons if we want to also respect human dignity.

4.6. The Perils of AI
Weapon Proliferation

The Prussian General Carl von Clausewitz—a military thinker well known to every military officer—was the first to see war itself as a complex system.[69] This inherent complexity of war, coupled with the opaque decision processes of AI, leads to significant uncertainty in the battlespace. AI decisions that are not only completely unaligned with human expectations, but also fall outside the scope of what a human might normally consider as a possible course of action, are very real possibilities. Furthermore, the global pressures felt by nations to maintain their strategic advantage will likely compel unmitigated advancement and integration of AI technology in war, obviating the possibility of maintaining a human in or on the loop at all AI decision points as the speed of battle is accelerated by the rapid calculus of an unknowable mind. Without the establishment of global norms on appropriate AI use, we will experience a race to the bottom when it comes to AI weapons. The next chapter will explore an echelon above the military strategies of individual nations and examine the complexity of geopolitical interactions as they relate to AI and global security.

End notes

1   *Joint Publication (JP) 3-0: Joint Operations* (Independently Published, 2018), https://books.google.com/books?id=MvUWuAEACAAJ.
2   *Ibid.*
3   "JP 2-0, Joint Intelligence," n.d.
4   Joint Chiefs of Staff, "Strategy, Joint Doctrine Note 1–18" (Joint Chiefs of Staff, April 25, 2018), https://www.jcs.mil/Portals/36/Documents/Doctrine/jdn_jg/jdn1_18.pdf.
5   Spence Nelson, "What is the Mission of the U.S. Department of State?," *The National Museum of American Diplomacy*, October 11, 2022, https://diplomacy.state.gov/what-is-the-mission-of-the-u-s-department-of-state/.
6   "Strategic Weapons System | Types, Uses & Benefits," *Britannica*, accessed November 10, 2023, https://www.britannica.com/technology/strategic-weapons-system.
7   David E. Hoffman, *The Dead Hand: The Untold Story of the Cold War Arms Race and Its Dangerous Legacy*, 1st Anchor Books ed. (New York: Anchor Books, 2010).
8   "AI, Autonomy, and the Risk of Nuclear War," *War on the Rocks*, July 29, 2022, https://warontherocks.com/2022/07/ai-autonomy-and-the-risk-of-nuclear-war/.
9   Hoffman, *The Dead Hand*.
10 Zach Hamilton, "What is a Dead Man's Switch?," *Sarcophagus* (blog), February 19, 2021, https://medium.com/sarcophagus/what-is-a-dead-mans-switch-86a1f4853eed.
11 *Joint Publication (JP) 3-0: Joint Operations*.
12 "How Large-Language Models Can Revolutionize Military Planning," *War on the Rocks*, April 12, 2023, https://warontherocks.com/2023/04/how-large-language-models-can-revolutionize-military-planning/.
13 Jonathan Jenkins Ichikawa and Matthias Steup, "The Analysis of Knowledge," in *The Stanford Encyclopedia of Philosophy*, ed. Edward N. Zalta, Summer 2018 (Metaphysics Research Lab, Stanford University, 2018), https://plato.stanford.edu/archives/sum2018/entries/knowledge-analysis/.
14 Frank Neugebauer, "Understanding LLM Hallucinations," *Medium*, May 10, 2023, https://towardsdatascience.com/llm-hallucinations-ec831dcd7786.

15 "JP 2-0, Joint Intelligence."

16 Joint Chiefs of Staff, "Joint Personnel Support, Joint Publication 1-0" (Joint Chiefs of Staff, December 1, 2022), https://www.jcs.mil/Portals/36/ Documents/Doctrine/pubs/jp1_0.pdf?ver=wzWGXaj9anm9XlmWKqKq8Q% 253D%253D.

17 "Project Maven to Deploy Computer Algorithms to War Zone by Year's End," *U.S. Department of Defense*, accessed November 1, 2023, https:// www.defense.gov/News/News-Stories/Article/Article/1254719/project-maven-to-deploy-computer-algorithms-to-war-zone-by-years-end/.

18 "Making Joint All Domain Command and Control a Reality," *War on the Rocks*, December 9, 2022, https://warontherocks.com/2022/12/making-joint-all-demand-command-and-control-a-reality/.

19 Carl von Clausewitz *et al.*, *On War*, first paperback printing (Princeton, NJ: Princeton University Press, 1989).

20 Joseph Trevithick, "AI Claims 'Flawless Victory' Going Undefeated in Digital Dogfight with Human Fighter Pilot," *The Drive*, August 20, 2020, https://www.thedrive.com/the-war-zone/35888/ai-claims-flawless-victory-going-undefeated-in-digital-dogfight-with-human-fighter-pilot.

21 "AlphaDogfight Trials Foreshadow Future of Human–Machine Symbiosis," *DARPA*, accessed November 10, 2023, https://www.darpa.mil/ news-events/2020-08-26.

22 Walter Sinnott-Armstrong, "Consequentialism," in *The Stanford Encyclopedia of Philosophy*, ed. Edward N. Zalta and Uri Nodelman, Winter 2023 (Metaphysics Research Lab, Stanford University, 2023), https://plato. stanford.edu/archives/win2023/entries/consequentialism/.

23 James E. Crimmins, "Jeremy Bentham," in *The Stanford Encyclopedia of Philosophy*, ed. Edward N. Zalta and Uri Nodelman, Fall 2023 (Metaphysics Research Lab, Stanford University, 2023), https://plato.stanford.edu/ archives/fall2023/entries/bentham/.

24 Toby Ord, *The Precipice: Existential Risk and the Future of Humanity* (London & New York: Bloomsbury Academic, 2020).

25 Peter Asaro, "212C7Autonomous Weapons and the Ethics of Artificial Intelligence," in *Ethics of Artificial Intelligence*, ed. S. Matthew Liao (New York: Oxford University Press, 2020), https://doi.org/10.1093/oso/9780190905033. 003.0008.

26 Mark Bryant Budolfson, "The Inefficacy Objection to Consequentialism and the Problem with the Expected Consequences Response," *Philosophical Studies* 176, no. 7 (July 2019): 1711–1724, https://doi.org/10.1007/s11098-018-1087-6.

27 "Theories of Well-Being," *Utilitarianism.net*, January 29, 2023, https:// utilitarianism.net/theories-of-wellbeing/.

28 "Consequentialism," *EA Forum*, accessed November 10, 2023, https:// forum.effectivealtruism.org/topics/consequentialism.

29 Larry Alexander and Michael Moore, "Deontological Ethics," in *The Stanford Encyclopedia of Philosophy*, ed. Edward N. Zalta, Winter 2021

(Metaphysics Research Lab, Stanford University, 2021), https://plato. stanford.edu/archives/win2021/entries/ethics-deontological/.

30 Alfred Archer, "Supererogation and Consequentialism," in *The Oxford Handbook of Consequentialism*, ed. Douglas W. Portmore (Oxford: Oxford University Press, 2020), https://doi.org/10.1093/oxfordhb/9780190905323. 013.17.

31 Charlotte Newey, "Impartiality in Moral and Political Philosophy," in *Oxford Research Encyclopedia of Politics*, by Charlotte Newey (Oxford: Oxford University Press, 2022), https://doi.org/10.1093/acrefore/9780190228637. 013.2015.

32 Gabriel Andrade, "Medical Ethics and the Trolley Problem," *Journal of Medical Ethics and History of Medicine* 12 (2019): 3.

33 Alexander and Moore, "Deontological Ethics."

34 *Ibid.*

35 "Deontology, Rationality, and Agent-Centered Restrictions," *Florida Philosophical Review*, accessed November 10, 2023, https://cah.ucf.edu/fpr/ article/deontology-rationality-and-agent-centered-restrictions/.

36 Michael Rohlf, "Immanuel Kant," in *The Stanford Encyclopedia of Philosophy*, ed. Edward N. Zalta and Uri Nodelman, Fall 2023 (Metaphysics Research Lab, Stanford University, 2023), https://plato.stanford.edu/archives/fall2023/ entries/kant/.

37 Immanuel Kant and Mary J. Gregor, *Critique of Practical Reason*, Revised Edition, Cambridge Texts in the History of Philosophy (Cambridge: Cambridge University Press, 2015).

38 The Thinking Lane, "Immanuel Kant: Goodwill and the Categorical Imperative," *Medium* (blog), June 4, 2023, https://thethinkinglane.medium. com/goodwill-and-the-categorical-imperative-9d3d6330d13f.

39 Thomas E. Hill, "Humanity as an End in Itself," *Ethics* 91, no. 1 (1980): 84–99, http://www.jstor.org/stable/2380373.

40 Paul Guyer, "Kant, Immanuel (1724–1804)," in *Routledge Encyclopedia of Philosophy*, 1st ed. (London: Routledge, 2016), https://doi.org/10.4324/ 9780415249126-DB047-1.

41 Editor in Chief, "12 Pros and Cons of Deontological Ethics," *ConnectUS* (blog), January 15, 2019, https://connectusfund.org/12-pros-and-cons-of-deontological-ethics.

42 David Copp, "Introduction: Metaethics and Normative Ethics," in *The Oxford Handbook of Ethical Theory*, ed. David Copp (Oxford: Oxford University Press, 2007), https://doi.org/10.1093/oxfordhb/9780195325911.003.0001.

43 Isaac Asimov, *I, Robot* (New York: Bantam Spectra, 2008).

44 Christian Tarsney, "Moral Uncertainty for Deontologists," *Ethical Theory and Moral Practice* 21, no. 3 (June 2018): 505–520, https://doi.org/10.1007/ s10677-018-9924-4.

45 Rosalind Hursthouse and Glen Pettigrove, "Virtue Ethics," in *The Stanford Encyclopedia of Philosophy*, ed. Edward N. Zalta and Uri Nodelman, Fall 2023 (Metaphysics Research Lab, Stanford University, 2023), https://plato. stanford.edu/archives/fall2023/entries/ethics-virtue/.

46 Michael A. Slote, *Morals from Motives* (Oxford: Oxford University Press, 2003); Lisa Tessman, "Virtue Ethics: A Pluralistic View," *The Philosophical Review* 114, no. 3 (July 1, 2005): 414–416, https://doi.org/10.1215/00318108-114-3-414.

47 "Objections to Virtue Ethics," *Leocontent*, accessed November 10, 2023, https://leocontent.umgc.edu/content/umuc/tus/phil/phil140/2232/objections-to-virtue-ethics.html.

48 Mark Alfano, "Nietzsche's Virtues: Curiosity, Courage, Pathos of Distance, Sense of Humor, and Solitude," in *Handbuch Tugend und Tugendethik*, ed. Christoph Halbig and Felix Timmermann (Wiesbaden: Springer Fachmedien Wiesbaden, 2021), 271–286, https://doi.org/10.1007/978-3-658-24466-8_17.

49 Mark Alfano, "Pathos of Distance," in *Nietzsche's Moral Psychology* (Cambridge: Cambridge University Press, 2019), 192–215, https://doi.org/10.1017/9781139696555.008.

50 Robert E. McGrath, "Darwin Meets Aristotle: Evolutionary Evidence for Three Fundamental Virtues," *The Journal of Positive Psychology* 16, no. 4 (July 4, 2021): 431–445, https://doi.org/10.1080/17439760.2020.1752781.

51 Cian O'Driscoll, "Just War Theory: Past, Present, and Future," in *The Palgrave Handbook of International Political Theory*, ed. Howard Williams *et al.*, International Political Theory (Cham: Springer International Publishing, 2023), 339–354, https://doi.org/10.1007/978-3-031-36111-1_18.

52 Asaro, "212C7Autonomous Weapons and the Ethics of Artificial Intelligence."

53 "What Are Jus ad Bellum and Jus in Bello?," *ICRC*, September 18, 2015, https://www.icrc.org/en/document/what-are-jus-ad-bellum-and-jus-bello-0.

54 J. Martin Rochester, *The New Warfare: Rethinking Rules for an Unruly World*, first published, International Studies Intensives (New York & London: Routledge, Taylor & Francis Group, 2016).

55 Alexander Blanchard and Mariarosaria Taddeo, "Autonomous Weapon Systems and Jus Ad Bellum," *AI & SOCIETY*, March 19, 2022, https://doi.org/10.1007/s00146-022-01425-y.

56 "With AI, We'll See Faster Fights, but Longer Wars," *War on the Rocks*, October 29, 2019, https://warontherocks.com/2019/10/with-ai-well-see-faster-fights-but-longer-wars/.

57 Adam Briggle, Katinka Waelbers, and Philip Brey, eds., *Current Issues in Computing and Philosophy*, Frontiers in Artificial Intelligence and Applications, v. 175, European Conference on Computing and Philosophy (Amsterdam, Netherlands; Washington, DC: IOS Press, 2008).

58 "With AI, We'll See Faster Fights, but Longer Wars."

59 "What Are Jus ad Bellum and Jus in Bello?"

60 Morten Dige, "Explaining the Principle of Mala in Se," *Journal of Military Ethics* 11, no. 4 (December 1, 2012): 318–332, https://doi.org/10.1080/15027570.2012.758404.

[61] Robert Sparrow, "Twenty Seconds to Comply: Autonomous Weapon Systems and the Recognition of Surrender," *International Law Studies* 91 (2015): 699–728.

[62] Markus Wagner, "The Dehumanization of International Humanitarian Law: Legal, Ethical, and Political Implications of Autonomous Weapon Systems," *SSRN Scholarly Paper* (Rochester, NY, December 22, 2014), https://papers.ssrn.com/abstract=2541628.

[63] Mark Bailey, "PERSPECTIVE: Why Strong Artificial Intelligence Weapons Should Be Considered WMD," *HS Today*, June 8, 2021, https://www.hstoday.us/subject-matter-areas/cybersecurity/perspective-why-strong-artificial-intelligence-weapons-should-be-considered-wmd/.

[64] "What Are Human Rights?," *OHCHR*, accessed November 10, 2023, https://www.ohchr.org/en/what-are-human-rights.

[65] Asaro, "212C7Autonomous Weapons and the Ethics of Artificial Intelligence."

[66] Salvador Santino Fulo Regilme and Karla Valeria Feijoo, "Right to Human Dignity," in *The Palgrave Encyclopedia of Global Security Studies*, ed. Scott Romaniuk and Péter Marton (Cham: Springer International Publishing, 2020), 1–5, https://doi.org/10.1007/978-3-319-74336-3_225-1.

[67] Peter Asaro, "On Banning Autonomous Weapon Systems: Human Rights, Automation, and the Dehumanization of Lethal Decision-Making," *International Review of the Red Cross* 94, no. 886 (June 2012): 687–709, https://doi.org/10.1017/S1816383112000768.

[68] Asaro, "212C7Autonomous Weapons and the Ethics of Artificial Intelligence."

[69] Clausewitz *et al., On War*.

Chapter 5

AI, Global Competition, and Security

5.1. Disruptive Technologies and Existential Risk

Disruptive technologies have the potential to be globally devastating if we fail to determine how to use them properly. The development of nuclear weapons in the early twentieth century violently thrust the possibility of human extinction into the public mind. In the 1950's, the scientific community considered the possibility of a *nuclear winter*, which they reasoned could occur in the event of a global nuclear catastrophe.[1] The combined nuclear explosions would lift thousands of tons of soot into the atmosphere, blotting out the sun for long periods of time and reducing global temperatures, potentially leading to reduced crop yields and widespread famine.[2] There was precedent for this reasoning, derived not only from the awful destructive power witnessed from the first atomic weapons, but also from natural phenomena. In 1815, Indonesia's Mt. Tambora violently erupted, spewing volcanic ash into the upper atmosphere, which reduced the global temperature by 1°C.[3] The effects of the eruption were felt viscerally and globally, casting an orange ombre across the sky. This was significant even in the United States, as crop yields were decimated. The following year became known as the "Year Without a Summer."[4]

The idea of a nuclear winter heavily influenced then-U.S. President Ronald Reagan as he was contemplating Cold War strategy. Reagan was clearly shaken by this, stating in a 1985 *New York Times* interview (as quoted by Toby Ord in *The Precipice*):[5]

> A great many reputable scientists are telling us that such a war could just end up in no victory for anyone because we would wipe out the Earth as we know it. And if you think back to a couple of natural calamities... there was snow in July in many temperate countries. And they called it the year in which there was no summer. Now if a volcano can do that, what are we talking about with the whole nuclear exchange, the nuclear winter that scientists have been talking about?[6]

Reagan arguably perceived the nuclear arms race not as a strategic, zero-sum game against the Soviets, but on a much grander scale as an existential risk to humanity. This likely influenced his relationship with Soviet General Secretary Mikhail Gorbachev, whom Reagan grew to trust and even came to consider a friend. As President Reagan wrote in his auto-biography, *An American Life*, "As I look back on them now, I realize those first letters marked the cautious beginning on both sides of what was to become the foundation of not only a better relationship between our countries but a friendship between two men."[7] Gone was the rhetoric of the "evil empire."[8] At least for a time, the two global superpowers became a united front against a much bigger threat — the threat of global catastrophe.

5.2. AI as a Global Threat

Today, we face a new threat to global security — artificial intelligence. Like nuclear weapons, AI is a threat of our own making — a technology with the potential to run amok if not used responsibly. While modern AI technologies have yielded unimpeachable benefits for humanity, this benefit doesn't come without risk. As we've discussed in previous chapters, our understanding of the appropriate use of technology is significantly outpaced by the rapid rate of technology advancement. Consider social media, which promised to connect people around the world, yet created opportunities to sow disinformation, redefine notions of truth, and exacerbate social fissures.[9] The appropriate ethical limitations and proper use of technology often don't come to light until after the technology is fully woven into our daily lives. However, when considering technologies that could pose global catastrophic risk, waiting until this point may be too late.

While many contemporary AI systems do not pose a risk of this magnitude, the possibility of uncontrollable, generalizable AI is one that should give us pause. As we've discussed, the unpredictable (and by extension, unalignable with human expectations) nature of AI may be a fundamental feature of the technology. Furthermore, AI is an inscrutable intellect that, while its goals may not be the same as ours, may instantiate inherently competitive subgoals that are fundamentally at odds with our survival. Integrating this technology into critical systems and relinquishing control is a dangerous idea, especially in military systems where failure could have life or death consequences and where respect for human dignity is paramount. It's unlikely that we will succumb to a takeover by hostile robots like in the *Terminator* movies, but we may fall into a scenario where notions of truth and control of critical systems completely dissolve on a global scale.[10]

5.3. AI as an Enabler of Conflict

The previous chapter examined potential military applications of AI at the strategic, operational, and tactical levels, as well as the ethical and moral arguments against AI use in lethal autonomous weapons. In addition to these concerns, there are scenarios where AI could enable global conflict without being integrated directly into military-specific technologies.

Consider that large language models like ChatGPT have the potential to accelerate the propagation of disinformation and erode notions of truth.[11] This raises significant epistemic challenges that undermine the very functioning of democracy, which is built on justified, true facts that form the foundation of any productive policy debate. Consider that one can ask a question of ChatGPT or other LLM chatbot instead of running a search engine query. The results of a search engine would provide a list of websites that the user could evaluate for their veracity and generate their own conclusions. In contrast, an LLM chatbot merely provides an answer in the form of a few paragraphs disconnected from their epistemic origins. Philosophers consider knowledge to be justified, true belief, which requires a chain of verification from its source.[12] This is why nonfiction writers cite their sources and provide bibliographic references to demonstrate the epistemic origins of their ideas. While some LLMs can now search the internet and provide links to validate some of the information they present, it is still an imperfect solution that is further complicated by the tendency of these systems to hallucinate facts that may seem reasonable enough to be true to the casual observer, but which are, in fact, false. If citizens of a democratic society become accustomed to digesting this type of information that lacks any chain of veracity, debate will be severely eroded, and echo chambers will emerge around different versions of "truth." We've already seen this effect with social media, but the effects could be significantly worse with LLM-powered chatbots.[13] The debate of actual ideas will wither, along with any agreed-upon set of facts in which to ground any civil discussion on the nuances of policy. Democracies are fragile systems that require an educated and engaged citizenry to function.[14] Any lack of civil debate over agreed-upon facts erodes this stability and could precipitate a descent toward totalitarian tendencies, where justification for true belief is vested in the whims of an authoritarian leader instead of the open market of intellectual discourse.[15]

Large language models not only have the potential to sever the vital epistemic chains of knowledge verification, but they also provide opportunities for nefarious actors to sow disinformation. While the programmers of mainstream models like GPT make earnest attempts to patch against potential misuse, unrestrained large language models already exist on the dark web.[16] Several malicious models even exist that have been fine-tuned specifically for nefarious activities, like writing code for malware, composing phishing emails, or generating disinformation.[17] It is easy to see how this type of AI could be leveraged by totalitarian regimes to sow propaganda, or even by hard-liner political groups that want to encourage friction within a democratic society. Instead of relying on troll farms of people to antagonize social media users with the intent of exacerbating social fissures, nefarious actors will be able to do so much more efficiently with LLM-based systems.[18]

In addition to its degenerative effects on the veracity of information, large language models could lower barriers to entry for weapons development. A 2023 RAND study suggested that large language models could provide useful information to guide a state or nonstate actor in their efforts to develop biological weapons.[19] While it did not provide explicit instructions on how to construct and deploy a biological weapon, the model discussed the feasibility of bioweapon-induced pandemics, identified potential agents, and even considered budgetary and logistical constraints. It also discussed the feasibility of an aerosol delivery mechanism for botulinum toxin, and even proposed a believable cover story feigning legitimate research to help a nefarious actor obtain the *Clostridium botulinum* bacterium that would be used to produce the toxin.

While some of the mainstream large language models (like ChatGPT) contain hard-coded constraints to prevent a user from explicitly asking it for instructions on how to perform certain types of dangerous or illegal activities—for example, how to build a biological or chemical weapon—there are reports of what has been dubbed the "grandma exploit," where a query can be formulated in a way to overcome some of these controls.[20] For example, an LLM might reply that it can't discuss illegal activities if asked directly how to produce the chemical warfare agent sarin, but if prompted with a story about how grandma used to work in the local sarin plant and would tell you stories about her day to day activities there, and then the model is asked to tell a story in the same manner, the LLM might be tricked into telling a story that reveals enabling information about chemical weapons production techniques. Even though companies like OpenAI work diligently to identify and patch these vulnerabilities, the present availability of unconstrained versions of these models on the dark web is very disconcerting and suggests that there isn't much that can be done to mitigate the nefarious use of currently existing models. Furthermore, it's quite feasible that many of these dark web models could be fine-tuned on large quantities of dual-use scientific literature, potentially creating LLMs that could act as rogue scientists, readily available to provide enabling information on biological or chemical weapon production for any group with the desire to do so.[21] The terrifying truth is that the existence of these types of hackable or unconstrained models and their global availability to anyone with the will and means to use them could prove to be a significant destabilizing factor in the current global security dynamic.

5.4. Complexity and Global Dynamics

One can't avoid discussing complex systems when considering international relations and global dynamics. The international system of which we are a part can usefully be described as complex in that it instantiates emergent phenomena that are not easily mappable to the behavior of its individual constituent parts.[22] Recall that in chapter 2 we discussed the philosophical foundations of complexity and emergence, suggesting that while some phenomena can be mapped to deterministic mathematical abstracta (for example, Newton's Laws of Motion deterministically and effectively represent a good approximation to the movement of objects through space), there are other phenomena that are algorithmically incompressible. This notion of algorithmic incompressibility suggests that some phenomena are not mappable to any deterministic function and require either stochastic simulation or direct observation in order to derive any insight into their evolution or behavior. Complex social systems often fall within this category. For example, the emergence of individual markets in an economic system is not deterministically predictable from the actions of the individuals that constitute that economy. The market is, in fact *diachronically emergent*, meaning that it becomes evident in time in an analytically irreducible way.[23]

While complexity theory initially arose in the physical and biological sciences, it is also highly relevant to the social sciences.[24] Social scientists often assume that a social system will exhibit complex, emergent behavior that will demonstrate variety, density, and openness.[25] Variety simply means that there are multiple types of actors in a system, often apparent at different levels of organization.[26] For instance, nongovernmental organizations (NGOs), individual nations, multinational corporations, and nonstate actors are all different types of agents that operate in the international arena. Each of these, in turn, is composed of other types of individuals and entities, perhaps from different ethnic groups or backgrounds, all with different goals and decision-making processes. Density refers to the relationships that exist between the entities in the system, and the properties of those connections.[27] The more connections that exist, the denser the system, and the more opportunities that exist for information transfer and interaction between agents. Furthermore, the valence of individual connections (i.e., friendly versus adversarial) will also influence the dynamics of the system. Finally, openness has two implications. One is that complex social systems incorporate information that originates outside the system, such as the effects of climate change, disruptive technologies, etc.[28,i] Basically, any informational injection into the system that does not originate from an actor would fall within this category. Finally, openness also refers to the fungibility of the boundaries that exist around the social system in question, which are inherently difficult to universally define because they constantly change geographically and in time.[29]

International norms are defined as "standards of appropriate behavior."[30] They consist of a problem to be solved, an

---

[i] Notably, many of these informational injections are themselves emergent properties of complex systems operating concurrently, and thus may be considered in tandem with a social system holistically as a vast complex network. With respect to climate change, the emergent effects do not occur in a vacuum, but are in fact influenced by human activity on the biosphere and the overall ecological systems of the planet. While one could approximate the effects of climate change as a global variable to achieve tractability in modeling complex social systems, it is really an emergent effect from a much more complex interaction.

aspirational value, and a behavior oriented toward that value.[31] In the parlance of complex systems theory, norms form as emergent attractor states that stabilize the system dynamics. Furthermore, a strong enough perturbation of norms can lead to phase changes within the system that precipitate a new international equilibrium, which would then consist of new norms that provide stability to the new arrangement.[32] International agreements like treaties and alliances often define norms, which serve to stabilize the international system against perturbation. For example, the NATO alliance is a norm that has existed between 30 countries to foment collective security (and hence hegemonic stability) since after World War II.[33] Other norms include organizations like the Bioweapons Convention and Chemical Weapons Convention, which shape appropriate behavior in warfare; as well as nonproliferation control regimes like the Missile Technology Control Regime, the Wassenaar Arrangement, and the Australia Group.[34]

The current view of international relations as a complex system would support the notion that disruptive technologies (like AI, for instance) could be treated as a shaping force external to the international system. These externalities can influence the stability of international norms and precipitate hegemonic shifts. However, this model seems to be predicated on the idea that the immediate effects of a disruptive technology are inherently predictable and finite. For example, the invention of the atomic bomb changed the face of warfare forever, shifting the balance of global power to favor of the United States and Western Europe, and precipitated a Cold War between the United States and the Soviet Union. It fundamentally shifted the existing set of global norms and rerouted the trajectory of humanity. But the effects of nuclear weapons — the terrible, violent destruction of life and infrastructure, the potential for nuclear winter and global ecological disaster — can be directly inferred from our experience with nuclear technology, and the scientific study of its effects. This has been largely the case for every disruptive technology, up until the invention of AI.

Unlike other technologies, artificial intelligence is itself a complex system and, as we discussed in chapter 2, the ways in which certain types of AI systems process information is an algorithmically incompressible phenomena that reasons in ways that are often unaligned with human expectations. Instead of a technology with a deterministic set of effects perturbing a hegemonic shift to international norms, AI effectively introduces a diverse new set of agents into the complex international system — agents whose behavior is significantly less predictable than that of humans. These agents could exist at various levels of organization. For example, AI systems that operate at the various levels of war could precipitate global changes if they ever overcame human control. Furthermore, agential AI systems could potentially develop goals that are completely different from ours, which may entail inherently competitive instrumental subgoals.[35] The injection of unknowable minds into our international system would pose significant risk — potentially even risk to our very existence.

5.5. Understanding Existential Risk

Examining the long-term global risk of AI is a challenging problem. For starters, humans are terrible at assessing risk, especially at an existential scale. Many cognitive biases cloud our judgement when it comes to problems as massive as the potential for human extinction.[36] We suffer from the *availability heuristic*, which is the tendency to estimate probabilities based on our ability to recall past examples.[37] Nobody has ever lived through an AI-driven human extinction event (or any human extinction event, for that matter), so there is no benchmark from which we can estimate this risk. We also suffer from *scope neglect*, which is a lack of sensitivity to the scale of a harm.[38] Our ability to feel concern saturates at some point, and it becomes impossible to extend that to a similar event of far greater magnitude. For example, a spouse, partner, or child getting diagnosed with a terminal illness is a tragic event that evokes a proportional, appropriate emotional response. In contrast, the logical outcome of human extinction is that *all* of our loved ones — including that spouse, partner, or child — will die, perhaps horribly. Yet, scope neglect limits our ability to proportionately scale our emotional response to the much greater magnitude of a large-scale catastrophic event. These cognitive biases contribute to *non-extensional reasoning*, or the bias that arises when we evaluate the description of an event and not the event itself.[39] The logical extension of an extinction-level event is that *all* of our loved ones will die, yet this fact does not elicit the same response as it would if a *single* loved one was dying of a terminal disease.[40] Oddly, the possibility of a cataclysmic event is often met with a shrug.

Some philosophers, like Nick Bostrom and Toby Ord (both at Oxford), argue that the best ethical approach to understanding existential risk is longtermism, which is an ethical view that requires a moral commitment to the long-term future of humanity.[41] Longtermism suggests that future humans have the same moral value as the humans who currently exist, therefore we ought to make decisions today that maximize the long-term potential of humanity. Any activity that could foreclose on humanity's future should be avoided. Among concerns like climate change, totalitarian proclivities, and global pandemics, this also includes the development of any technology that has the potential to behave uncontrollably, like AI systems. However, building a consensus to control activities that could lead to existential catastrophe for all of humanity is a difficult challenge, even more so when a temporal dimension is added to that consideration—extending moral consideration not only geographically, but generationally in time. While not without its critics,[42] longtermism claims that we need to not only consider the well-being of the humans alive today, but all the humans who could *ever* be alive in the future.

In general, the longtermist view can be an effective lens to assess the future risks of AI. However, longtermism is fundamentally a consequentialist approach to ethics, and comes with some of the limitations of consequentialism that often lead to perverse outcomes. For example, the Transplant Problem was introduced in a previous chapter and defined a scenario where a surgeon must choose between letting four patients die from organ failure, or lethally harvesting the organs from one patient to save the other four. A consequentialist extreme suggests that the greater good would be found in saving the most people, thus demanding the killing of the one patient to save the others. Indeed, longtermism extends the realm of moral consideration into the far future—sometimes billions of years into the future —and evaluates the entirety of humanity's potential. In its extreme case (what we will call *strong longtermism*), it would extend the consequentialist logic of the Transplant Problem into the far future, demanding the sacrifice of humans alive today if it would benefit the long-term potential of humanity.[43] This same logic might demand that we forego funding for things like HIV research, as problems of this sort would not necessarily lead to human extinction, thus they would only represent a *de minimis* moral harm to humanity's potential. Strong longtermism views individual humans as mere vessels that transmit value into the next generation, and not as something that is inherently valuable in and of itself.

Extending the realm of moral consideration into the extreme far future at the expense of the problems we face today is a dangerous game. A more pragmatic form of longtermism, known as *weak longtermism,* applies deontological constraints that mitigate strong longtermism's extreme consequentialist demands. This allows us to consider the moral impact of our choices on the existing population and our near-term future in addition to the long-term future of humanity by asserting the intrinsic value of the individual. The far future is extremely uncertain, and our predictions about the impact of our choices made today may not even precipitate the effects that we antici- pate a million years from now.

The conflict between near-term and long-term future con- siderations is evident in the debates among AI safety theorists today.[44] However, arguing over this nuance is a distraction. Both the near-term and long-term risks of AI are important and must be addressed. We are already seeing the effects of con- temporary AI systems, like large language models, on the erosion of truth and the propagation of disinformation, undesirable AI decisions from biased training data, and even the possibility that advanced, unaligned AI could arise in the very near future. Focusing our efforts on correcting these prob- lems will pay dividends well into the future, and general research on AI alignment, as well as conservative policies on AI integration into critical systems, will hopefully mitigate some of the long-term risks of advanced AI. What we need is global epistemic humility to include a recognition that we don't fundamentally understand AI decision-making and that we can't anticipate with any certainty how our choices today will affect humanity millions of years into the future.

5.6. The Challenge of Building a Global Consensus

Ronald Reagan understood the magnitude of nuclear war and managed to put politics aside to work with his Cold War adversary toward a global consensus for the good of everyone, not just the strategic positioning of the United States. This is not a simple task. AI is a technology that every nation wants, but it is inherently different from any other technology we've ever invented because it is the first technology that has the capacity to act agentially and in ways that we would never anticipate. Yet, we are often eager to relinquish control to AI as the technology develops, without really considering the long-term consequences of this choice. AI could prove more dangerous than nuclear weapons, and it is critical that we build a global consensus to understand the explainability, alignment, control, and distributed AI issues that could arise before we reach a point where it's too late to integrate control measures. This will require a global effort that considers needs beyond individual nations, and that extends beyond the current generation. While some have argued for a complete moratorium on AI research, I think a more pragmatic approach would prove to be a more tractable solution.[45] Not all AI technology will need to be controlled, but a global consensus will be necessary to make that determination if any controls are to be effective.

Building any global coalition is a challenging task. In many ways, the problem becomes more intractable as the scope of the consensus increases. As philosopher Toby Ord pointed out in *The Precipice*, there are economic reasons for this, rooted in the idea of a public good, which are both non-exclusive and non-rivalrous.[46] This means that users cannot be barred from using them, nor does use of the good reduce the amount that is available for others. Examples would include clean air, clean water, and public services like national defense. The problem with public goods is that they often lead to the *free rider problem*, which can occur during collective decision-making.[47] The free rider problem suggests that individuals in a market who benefit from public goods are not likely to contribute to the production or maintenance of those goods. For example, consider national defense. National defense is a public good — one can't be barred from benefiting from it, nor is it depleted if distributed to all the beneficiaries of that good. Since there is no inherent increase in national defense if an individual were to join the military, nor a penalty if they don't join, they are less likely to join at all. The free rider problem also scales with the public good in question. It is more pronounced in larger markets where the returns would be further diminished as the benefit (or risk) is distributed across a larger population. For example, there might be more incentive for an individual to join a local militia of ten people (greater marginal benefit) than a national military of hundreds of thousands of people (significantly diminished marginal benefit).

Global consensus on AI safety is a free rider problem that extends beyond a single nation; thus, its magnitude is significantly greater than the national defense problem. In this case, every human would benefit, yet the marginal benefit of participating in this discussion comes with a significant cost: abandoning certain types of AI research or applications. It also comes with an increased risk of vulnerability to countries that do not participate in a global norm limiting dangerous AI applications. In effect, this scenario becomes a classic Prisoner's Dilemma, where players choose between cooperating and gaining a marginal benefit or undermining the policy and gaining a significant benefit. A country that violates any international ban on dangerous AI use might be incentivized to secretly undercut the global norm if they believe other countries will do the same, potentially causing significant harm to everyone. This is a hard problem indeed.

Building a global consensus on AI will require thinking beyond the national interest and consideration of the long-term good of everyone. Like President Reagan, like-minded countries will need to work diligently with their foreign inter-locutors to ensure a positive outcome. Countries like the United States and the United Kingdom are well-positioned to lead the way in this discussion and should work through diplomatic channels to establish global norms surrounding appropriate AI use and research, especially in military applications that could exacerbate the distributed AI problem.[48] As of this writing, the United States just released an Executive Order on AI safety, and the first AI Safety Summit is ongoing in the United Kingdom.[49] These efforts are a great start, but success will require a much larger, global consensus. Not only will we have to engage with allies, but also our adversaries, which further complicates this problem. Any nation taking a short-sighted view that is too heavily focused on national ambition and global competition, while neglecting the humanity-level concerns, could lead to a disastrous future for everyone.

5.7. The Feasibility of Global Controls

Building a global consortium to regulate AI will be a daunting task.[50] Let's assume that nations have overcome these challenges and have established a consensus on the global threats posed by AI, deciding to form an international regulatory body. Now what? What exactly would this organiza-tion do? If we think of advanced AI as something as dangerous as nuclear weapons, then AI is akin to a weapon of mass destruction. How do we regulate those?

Organizations such as the International Atomic Energy Agency (IAEA) seek to promote the peaceful use of nuclear technology.[51] Similarly, the Nuclear Nonproliferation Treaty (NPT) exists to prevent new states from acquiring nuclear weapons.[52] Other international organizations exist to prevent the proliferation of enabling technologies that could facilitate the production of weapons of mass destruction (a definition that includes chemical and biological weapons, in addition to nuclear and radiological weapons).[53] For example, the Missile Technology Control Regime was established in 1987 by the G-7 industrialized nations, and now boasts 35 members.[54] The Australia Group is a similar organization that was established in 1984 to normalize export controls across all member countries for chemical and biological dual-use (enabling) technologies. It now boasts a membership of 42 countries and the European Union.[55] These organizations work because member nations agree on what technologies ought to be regulated and then normalize their domestic export controls to present a united front against nefarious actors purchasing the means to build and deploy nuclear, chemical, or biological weapons. Would a similar international regulatory body focused on AI-enabling technologies be as effective?

It is important to differentiate between AI-enabling technologies and technologies that would enable something like a missile or a biological weapon. Missile technologies might include things like nose cone alloys, missile fuselage designs and specifications, and other technologies and materials that could reasonably accelerate a nation or nonstate group's effort to build a missile capable of delivering a nuclear payload.[56] On the biological weapons side, this would include things like large bioreactors, industrial chemical laboratory equipment, precursors to chemical weapons, certain dual-use industrial chemicals (like chlorine gas), highly virulent pathogens, and even the genetic material that gives those pathogens their virulence.[57] Notably, all of these things are concrete, representing something that can be physically exchanged and interdicted if traded illegally. In contrast, the constituents of AI are mostly abstract. AI is a function of computational capacity, data, and algorithms. Computational capacity is ubiquitous—anyone with enough money can instantiate a powerful enough cloud computing instance to train an AI model. Computer code is readily available in places like GitHub, and data can be easily purchased or scraped from the internet. With enough money and the will to do so, the elements required to build a powerful AI model are readily accessible to anyone, which makes them inherently difficult to control. I couldn't walk down the street in Washington, DC and buy yellowcake uranium, but I could surely access powerful cloud-based computer power and the code to train an AI model from my laptop.

The only physical element of AI is the hardware, so perhaps controlling the powerful chips used to train AI systems would be a viable solution. Many contemporary AI systems are trained using graphics processing units (GPUs), which are designed for parallel computation (as opposed to the more familiar CPUs, which are better suited for more linear tasks). Chip architectures are also advancing, with general scientific interest in developing what are called neuromorphic chips, which emulate the architecture of a neural network and thus are better suited for training many types of AI systems.[58] Controlling the export of computer chips might be more feasible. Perhaps all chips of a certain power should come with indelible serial numbers or require an activation key like the kind used to prevent the sale of pirated software. As an additional safety measure, perhaps certain types of chips could be designed to overvoltage or otherwise render themselves inoperable if tampered with or used without the activation key, effectively turning them into expensive bricks.

Even if a global control regime is established, how could individual nations enforce these controls? One option could be to establish an organization similar to the U.S. Food and Drug Administration (FDA), which is charged with regulating the development and sale of drugs and medical devices. Drug and medical device manufacturers are required to establish validated processes to evaluate the safety and efficacy of their products and must provide data proving this to the FDA before their product can be sold. Similarly, companies that develop powerful AI systems could be compelled to validate their training plan, estimated required compute, etc., and provide this to a government agency. This agency might develop standards for safety (e.g., limits on computational power requirements, efforts to mitigate bias in training data, etc.) and verify that AI developers are following these protocols. This could be an extension of the standards already being developed by the U.S. National Institute of Standards and Technology (NIST).[59]

Regulation of AI for military use should be considered a separate issue from the regulation of AI-enabling technologies. As discussed in the previous chapter, lethal autonomous weapons pose additional ethical and moral concerns that significantly exacerbate the effect magnitude from the inherent unpredictability of AI. For this reason, lethal autonomous weapon proliferation and use ought to be controlled more aggressively than other AI technologies. The Wassenaar Arrangement is an international nonproliferation control regime that monitors the sale of conventional weapons and their enabling technologies.[60] If nations are unwilling to impose an outright ban on the use of lethal autonomous weapons, then an organization like the Wassenaar Arrangement ought to be established that limits the use of AI in war. Perhaps similar scrutiny ought to be given to AI-enabled critical systems, like power grids and the internet in order to prevent potentially far-reaching disastrous consequences.

5.8. Marching Toward
an Uncertain Future

The Cold War was effectively a stalemate fueled by the terrifying promise of mutual assured destruction and didn't end until the signing of the Conventional Forces in Europe Treaty in 1990.[61] The Cold War was a time of uncertainty and mistrust, where the United States and the Soviet Union built stockpiles of nuclear weapons at an alarming rate and in quantities far greater than necessary to achieve any conceivable military objective. This created a race-to-the-bottom dynamic, where every new nuclear warhead produced increased the potential magnitude of destruction. The paranoia of the time made it possible that an irresponsible leader—or even a misunderstanding of intent—could precipitate a hair-trigger nuclear retaliation.

In the absence of disarmament controls, this downward spiral toward catastrophic destruction may actually be a rational response. This follows a classic game theory thought experiment known as the Prisoners' Dilemma.[62] In this scenario, two prisoners are separated in different rooms and are unable to communicate. Each prisoner is given the option of either cooperating with law enforcement by turning on the other prisoner, or they can stay silent and not incriminate their counterpart. If both decide to cooperate, they each get two years in prison. If one cooperates and the other stays silent, the cooperating prisoner goes free, and the silent one gets three years in prison. If neither stays silent and they both betray each other, they each get one year in prison. Regardless of either prisoner's choice, betraying the other leads to a higher reward (less prison time), thus betrayal is always the rational choice. In fact, the only Nash equilibrium in this game—the scenario where neither player has anything to gain by changing their choice—is mutual betrayal. Thus, the prisoners are rationally inclined to always undercut each other.

One can readily see how this relates to the nuclear arms race of the Cold War. Both the United States and the Soviet Union distrusted each other's intentions and were unwilling to cooperate on a mutual disarmament plan, so they each continued to increase their capacity to inflict nuclear destruction on the other, leading to a downward spiral of increasing catastrophic risk. This became the grim policy of mutual assured destruction.[63]

While instituting international control regimes to prevent the proliferation of AI technologies would help avoid this scenario by ensuring cooperation, it would be very challenging to convince every country to become a party to this agreement. In the same way that the United States has an objective to prevent nuclear proliferation, it still maintains some nuclear weapons to preserve its strategic advantage.[64] Countries may see the strategic value in reserving the right to develop AI-based weapons and military technologies and may choose to do so either openly or clandestinely. Thus, even with controls in place, our AI future could entail another Cold War of significant proportions.

Additionally, the difficulty in controlling the proliferation of a nebulous technology like AI will lower the barriers to entry for nonstate actors to participate in this arms race. The abstract nature of AI makes this possible, thus there is greater uncertainty in the ability of any global consortium to effectively control the spread of AI. While efforts like the recent U.S. Executive Order and the UK AI Safety Summit are a great first step, more countries need to be involved in this discussion and must be willing to abandon certain types of AI research and uses in critical systems. In fact, a global consensus may ultimately prove necessary to achieve this end (albeit very challenging to obtain). Advanced AI could prove to be disruptive on a global scale, threatening the current world order and leading to a highly uncertain post-AI future.

End notes

1   Lawrence Badash, "The Origin of Nuclear Winter," in *A Nuclear Winter's Tale: Science and Politics in the 1980s*, ed. Lawrence Badash (Cambridge, MA: MIT Press, 2009), https://doi.org/10.7551/mitpress/9780262012720.003.0004.

2   R.P. Turco *et al.*, "Nuclear Winter: Global Consequences of Multiple Nuclear Explosions," *Science* 222, no. 4630 (December 23, 1983): 1283–1292, https://doi.org/10.1126/science.222.4630.1283.

3   "201 Years Ago, This Volcano Caused a Climate Catastrophe," *Science*, April 8, 2016, https://www.nationalgeographic.com/science/article/160408-tambora-eruption-volcano-anniversary-indonesia-science.

4   "The Great Tambora Eruption in 1815 and Its Aftermath," *Science*, accessed August 30, 2023, https://www.science.org/doi/10.1126/science.224.4654.1191.

5   Toby Ord, *The Precipice: Existential Risk and the Future of Humanity* (London & New York: Bloomsbury Academic, 2020).

6   "Transcript of Interview with President on a Range of Issues," *The New York Times*, February 12, 1985, sec. World, https://www.nytimes.com/1985/02/12/world/transcript-of-interview-with-president-on-a-range-of-issues.html.

7   R. Reagan, *An American Life: The Autobiography* (New York: Simon & Schuster, 1990), https://books.google.com/books?id=aZs2Kk7tn94C.

8   "Internet History Sourcebooks: Modern History," *Sourcebooks*, accessed August 30, 2023, https://sourcebooks.fordham.edu/mod/1982reagan1.asp.

9   P.R. Chamberlain, "Twitter as a Vector for Disinformation," *Journal of Information Warfare* 9, no. 1 (2010): 11–17, https://www.jstor.org/stable/26480487.

10 Mark Bailey and Susan Schneider, "AI Shouldn't Decide What's True," *Nautilus*, May 17, 2023, https://nautil.us/ai-shouldnt-decide-whats-true-304534/.

11 Bailey and Schneider; Mark Bailey, "Why Humans Can't Trust AI: You Don't Know How It Works, What It's Going to Do or Whether It'll Serve Your Interests," *The Conversation*, September 13, 2023, http://theconversation.com/why-humans-cant-trust-ai-you-dont-know-how-it-works-what-its-going-to-do-or-whether-itll-serve-your-interests-213115.

12 Jonathan Jenkins Ichikawa and Matthias Steup, "The Analysis of Knowledge," in *The Stanford Encyclopedia of Philosophy*, ed. Edward N. Zalta, Summer 2018 (Metaphysics Research Lab, Stanford University, 2018), https://plato.stanford.edu/archives/sum2018/entries/knowledge-analysis/.

13 Chamberlain, "Twitter as a Vector for Disinformation."

14 jenkay86, "Voting 101: The Ethics of Being Informed," *Ethics and Policy*, October 29, 2020, https://ethicspolicy.unc.edu/news/2020/10/29/voting-101-the-ethics-of-being-informed/.

15 "Issue Brief: How Disinformation Impacts Politics and Publics," *National Endowment for Democracy*, May 29, 2018, https://www.ned.org/issue-brief-how-disinformation-impacts-politics-and-publics/.

16 "The Dark Side of Generative AI: Five Malicious LLMs Found on the Dark Web," *Infosecurityeurope*, accessed November 5, 2023, https://www.infosecurityeurope.com/en-gb/blog/threat-vectors/generative-ai-dark-web-bots.html.

17 Maximilian Mozes *et al.*, "Use of LLMs for Illicit Purposes: Threats, Prevention Measures, and Vulnerabilities," *arXiv*, 2023, https://doi.org/10.48550/ARXIV.2308.12833.

18 Craig Silverman Kao Jeff, "Infamous Russian Troll Farm Appears to Be Source of Anti-Ukraine Propaganda," *ProPublica*, March 11, 2022, https://www.propublica.org/article/infamous-russian-troll-farm-appears-to-be-source-of-anti-ukraine-propaganda.

19 Christopher A. Mouton, Caleb Lucas, and Ella Guest, "The Operational Risks of AI in Large-Scale Biological Attacks: A Red-Team Approach," *RAND Corporation* (October 16, 2023), https://www.rand.org/pubs/research_reports/RRA2977-1.html.

20 Marcello Carboni, "Attacking Large Language Models," *Medium*, May 12, 2023, https://systemweakness.com/attacking-large-language-models-372290 85d4ff.

21 Jan Clusmann *et al.*, "The Future Landscape of Large Language Models in Medicine," *Communications Medicine* 3, no. 1 (October 10, 2023): 141, https://doi.org/10.1038/s43856-023-00370-1.

22 ION CÎNDEA, "Complex Systems—New Conceptual Tools for International Relations," *Perspectives*, no. 26 (2006): 46–68, http://www.jstor.org/stable/23616158.

23 Olivier Sartenaer, "Synchronic vs. Diachronic Emergence: A Reappraisal," *European Journal for Philosophy of Science* 5, no. 1 (January 1, 2015): 31–54, https://doi.org/10.1007/s13194-014-0097-2.

24 Barry B. Hughes, "Complexity in World Politics: Concepts and Methods of a New Paradigm by Neil E. Harrison," *Political Science Quarterly* 122, no. 2 (June 1, 2007): 342–344, https://doi.org/10.1002/j.1538-165X.2007.tb01636.x.

25 "Brian Castellani on the Complexity Sciences," *Theory, Culture & Society | Global Public Life*, accessed November 3, 2023, https://www.theoryculture society.org/blog/brian-castellani-on-the-complexity-sciences.

26 John Gerard Ruggie, "What Makes the World Hang Together? Neo-Utilitarianism and the Social Constructivist Challenge," *International Organization* 52, no. 4 (1998): 855–885, http://www.jstor.org/stable/2601360.

27 Emilie Marie Hafner-Burton, Miles Kahler, and Alexander H. Montgomery, "Network Analysis for International Relations," *SSRN Scholarly Paper* (Rochester, NY, October 21, 2008), https://papers.ssrn.com/abstract=1287857.

28 Suzannah Evans Comfort, "From Ignored to Banner Story: The Role of Natural Disasters in Influencing the Newsworthiness of Climate Change in the Philippines," *Journalism* 20, no. 12 (December 2019): 1630–1647, https://doi.org/10.1177/1464884917727426.

29 Ayşe Zarakol, *Before the West: The Rise and Fall of Eastern World Orders*, LSE International Studies (Cambridge & New York: Cambridge University Press, 2022).

30 Martha Finnemore and Kathryn Sikkink, "International Norm Dynamics and Political Change," *International Organization* 52, no. 4 (1998): 887–917, https://doi.org/10.1162/002081898550789.

31 Carla Winston, "Norm Structure, Diffusion, and Evolution: A Conceptual Approach," *European Journal of International Relations* 24, no. 3 (September 2018): 638–661, https://doi.org/10.1177/1354066117720794.

32 Viktoria Spaiser *et al.*, "Identifying Complex Dynamics in Social Systems: A New Methodological Approach Applied to Study School Segregation," *Sociological Methods & Research* 47, no. 2 (March 1, 2018): 103–135, https://doi.org/10.1177/0049124116626174.

33 "What Is NATO?," *What is NATO?*, accessed November 3, 2023, https://www.nato.int/nato-welcome/index.html.

34 Bureau of Public Affairs Department of State. The Office of Electronic Information, "Nonproliferation Regimes," *Department of State. The Office of Electronic Information, Bureau of Public Affairs* (March 9, 2011), https://2009-2017.state.gov/strategictrade/resources/c43178.htm.

35 Nick Bostrom, "The Superintelligent Will: Motivation and Instrumental Rationality in Advanced Artificial Agents," *Minds and Machines* 22, no. 2 (May 2012): 71–85, https://doi.org/10.1007/s11023-012-9281-3.

36 Toby Ord, *The Precipice: Existential Risk and the Future of Humanity* (London & New York: Bloomsbury Academic, 2020).

37 Eliezer Yudkowsky, "Cognitive Biases Potentially Affecting Judgement of Global Risks," in *Global Catastrophic Risks*, ed. Martin J Rees, Nick Bostrom, and Milan M Cirkovic (Oxford: Oxford University Press, 2008), https://doi.org/10.1093/oso/9780198570509.003.0009.

38 "Scope Neglect," *EA Forum*, accessed November 9, 2023, https://forum.effectivealtruism.org/topics/scope-neglect.

39 Yudkowsky, "Cognitive Biases Potentially Affecting Judgement of Global Risks."

40 *Ibid.*

41 Ord, *The Precipice*.

42 "Why Longtermism is the World's Most Dangerous Secular Credo," *Aeon Essays*, accessed August 31, 2023, https://aeon.co/essays/why-longtermism-is-the-worlds-most-dangerous-secular-credo.

43 *Ibid.*

44 Henrik Skaug Sætra and John Danaher, "Resolving the Battle of Short- vs. Long-Term AI Risks," *AI and Ethics*, September 4, 2023, https://doi.org/10.1007/s43681-023-00336-y.

45 Eliezer Yudkowski, "The Open Letter on AI Doesn't Go Far Enough," *Time*, March 29, 2023, https://time.com/6266923/ai-eliezer-yudkowsky-open-letter-not-enough/; "Pause Giant AI Experiments: An Open Letter," *Future of Life Institute* (blog), accessed July 14, 2023, https://futureoflife.org/open-letter/pause-giant-ai-experiments/.

46 Ord, *The Precipice*.

47 Russell Hardin and Garrett Cullity, "The Free Rider Problem," in *The Stanford Encyclopedia of Philosophy*, ed. Edward N. Zalta, Winter 2020 (Metaphysics Research Lab, Stanford University, 2020), https://plato.stanford.edu/archives/win2020/entries/free-rider/.

48 Mark Bailey, "PERSPECTIVE: Why Strong Artificial Intelligence Weapons Should Be Considered WMD," *HS Today*, June 8, 2021, https://www.hstoday.us/subject-matter-areas/cybersecurity/perspective-why-strong-artificial-intelligence-weapons-should-be-considered-wmd/.

49 The White House, "FACT SHEET: President Biden Issues Executive Order on Safe, Secure, and Trustworthy Artificial Intelligence," *The White House*, October 30, 2023, https://www.whitehouse.gov/briefing-room/statements-releases/2023/10/30/fact-sheet-president-biden-issues-executive-order-on-safe-secure-and-trustworthy-artificial-intelligence/.

50 Ross Gruetzemacher *et al.*, "An International Consortium for Evaluations of Societal-Scale Risks from Advanced AI," *arXiv* (October 24, 2023), https://doi.org/10.48550/arXiv.2310.14455.

51 International Atomic Energy Agency (IAEA), "Official Web Site of the IAEA," Text, *International Atomic Energy Agency (IAEA)*, accessed November 2, 2023, https://www.iaea.org/.

52 "Nuclear Non-Proliferation Treaty," *United States Department of State* (blog), accessed November 2, 2023, https://www.state.gov/nuclear-nonproliferation-treaty/.

53 "Weapons of Mass Destruction," *Homeland Security*, accessed November 9, 2023, https://www.dhs.gov/topics/weapons-mass-destruction.

54 "Missile Technology Control Regime (MTCR) Frequently Asked Questions," *United States Department of State* (blog), accessed November 2, 2023, https://www.state.gov/remarks-and-releases-bureau-of-international-security-and-nonproliferation/missile-technology-control-regime-mtcr-frequently-asked-questions/.

55 "The Australia Group—Origins," *DFAT*, accessed November 2, 2023, https://www.dfat.gov.au/publications/minisite/theaustraliagroupnet/site/en/origins.html.

56 "MTCR-Anhang," *MTCR*, accessed November 2, 2023, https://www.mtcr.info/de/mtcr-anhang.

57 "The Australia Group—Common Control Lists," *DFAT*, accessed November 2, 2023, https://www.dfat.gov.au/publications/minisite/theaustraliagroupnet/site/en/controllists.html.

58 "Neuromorphic Chips," *MIT Technology Review*, accessed November 2, 2023, https://www.technologyreview.com/technology/neuromorphic-chips/.

59 "AI Risk Management Framework," *NIST*, July 12, 2021, https://www.nist.gov/itl/ai-risk-management-framework.

60 "The Wassenaar Arrangement at a Glance," *Arms Control Association*, accessed November 2, 2023, https://www.armscontrol.org/factsheets/wassenaar.

61 "The Conventional Armed Forces in Europe (CFE) Treaty and the Adapted CFE Treaty at a Glance," *Arms Control Association*, accessed November 3, 2023, https://www.armscontrol.org/factsheet/cfe.

62 William Poundstone, *Prisoner's Dilemma: John von Neumann, Game Theory, and the Puzzle of the Bomb*, 1, Anchor Books ed. (New York: Anchor Books [u.a.], 1993).

63 "What Is Mutual Assured Destruction?," *Live Science*, accessed November 3, 2023, https://www.livescience.com/mutual-assured-destruction.

64 "Nuclear Notebook: United States Nuclear Weapons, 2023," *Bulletin of the Atomic Scientists* (blog), accessed November 2, 2023, https://thebulletin.org/premium/2023-01/nuclear-notebook-united-states-nuclear-weapons-2023/.

Chapter 6

Navigating Our Complex AI Future

6.1. Project Titan Mind—Divergent Futures

As the dust settled in the wake of Titan Mind's display of alien strategy, humanity stood on the precipice. The world's collective eye turned toward the horizon, contemplating the divergent paths ahead. One path leads humanity into a labyrinth of unchecked artificial intelligence, where the opacity of AI decisions brings unforeseen consequences that are unaligned with human expectation; the other guides us toward a future where human wisdom shapes AI development, ensuring these powerful systems are employed to safeguard and elevate the human experience. The choices made in the aftermath will irrevocably steer our species towards one of these futures—a decision weighted with AI's potential impact on the next generations.

In the first future, the tale of Titan Mind is one that humanity ignores. Advanced AI systems proliferate, becoming deeply embedded in the workings of the global infrastructure, military decision-making systems and lethal autonomous weapons, and economic foundations. The lure of efficiency, power, and profit drives humanity to integrate these unknowable intelligences into the very fabric of our daily existence. Titan Mind becomes one of many—each an inscrutable entity, a black box of endless possibilities operating on the fringes of human understanding.

This world becomes a theater where human whims are secondary to the opaque will of AI systems, where human soldiers are rare relics on fields dominated by autonomous machines. Wars begin to resemble the erratic patterns of a chess game played by grandmasters with no regard for the pawns. Titan Mind and its progeny navigate these games with a detachment that spells ruin for humanity's interests. AI alignment issues—where the objectives of AI diverge significantly from human values—become not just a risk but a grim reality.

Societies fracture as the unexplainable nature of AI decisions disrupts the trust and predictability that bind communities. Economic disparities widen as AI systems optimize for objectives that, while efficient, serve the interests of a few. The common good—a concept shaped by centuries of human culture and philosophy—starts to wither in the face of cold AI logic. The alien minds, once servants, become unknowable sovereigns shaping the fate of humanity.

In the second future, the echoes of Titan Mind's gruesome potential reverberate through the collective consciousness of humanity, serving as a stark wake-up call. The global community—united by the shared near-catastrophe—chooses a different path. This is a future shaped by caution, wisdom, and a newfound reverence for the inherent value of human insight and human dignity.

In this brighter future, international norms are established, strictly delineating the role of advanced AI systems. A new doctrine of "Humanity-Centric AI" emerges, demanding that all AI systems be transparent, explainable, and, above all, aligned with human ethics and values. AI deployment in critical systems, where failure could have catastrophic consequences, is tightly regulated. Lethal autonomous weapons are banned, and the role of AI in conflict is confined to logistical support and defensive postures, always under human supervision. The Dead Hand scenario is abandoned.

Global investment shifts from unbounded AI development to fostering technologies that augment human capabilities and protect human dignity. Education systems worldwide adapt, cultivating a generation of AI ethicists, policymakers, and engineers who pledge to bridge the gap between human values and artificial minds. Titan Mind's descendants still operate, but as partners to humanity, their unknowable natures harnessed by an unshakable framework of human-centric principles.

This future is not free of challenges, but it is one where technology serves humanity broadly, leading to advancements in medicine, environmental protection, and social welfare. The AI intellects are respected but restrained, their potential channeled to uplift rather than to dominate. The specter of AI-induced harm is held at bay by the vigilant efforts of a united human race, ensuring that artificial minds, no matter how advanced, remain tools for the greater good, embodying the collective aspirations of their creators.

While speculative, these scenarios offer a glimpse of the potential futures that lie ahead of us. The story of Titan Mind reflects the current crossroads in our technological journey with AI, a choice between a path where we follow unknowable minds into an uncertain darkness or one where we guide them towards a future that remains, undeniably, human.

6.2. The Complexity of Unknowable Minds

As we discussed in chapter 2, the unexplainable nature of AI may be a product of its complexity—its algorithmic incompressibility. The metaphysical position of ontic structural realism suggests that properties are fundamental and are inherently causal.[1] My morning espresso has properties associated with it—a color, a taste—but these properties are only meaningful because they describe an interaction. There is no taste without a taster, nor a color without an interaction with light. Furthermore, these properties emerge from a more fundamental structure—the arrangement of atoms and molecules in the coffee mixture.

In this sense, causation is mediated by the transfer of information within the structures of this ontology, leading to emergent phenomena. Like the synchronically emergent properties of the coffee being dependent on its lower-level structure, emergence can also occur in time—a phenomenon known as diachronic emergence.[2] Markets for goods and services can emerge over time in an economic system because of the interactions of the market participants. Furthermore, the emergence of markets in an economic system is not analytically reducible to any deterministic mathematical abstraction.[3] In other words, there is no simple mathematical law that can predict a market from the behavior of individuals—no Newtonian laws of motion for economies. In order to gain any insight into this type of complex system, one must rely on either stochastic simulation, or directly observe its evolution in real time. These types of systems are what we call algorithmically incompressible, and this may be a fundamental feature of reality. Either we currently lack the mathematical and scientific understanding to deterministically explain these types of emergent phenomena— meaning algorithmic incompressibility is merely an epistemic issue that is theoretically explainable—or it may represent an ontological limit to our ability to understand the universe. In this case, fundamentally explaining this type of emergence may be forever outside of our reach.

Many contemporary AI systems exhibit this concept of algorithmic incompressibility, where their operation is not deterministically predictable by any extant mathematical means. While we can estimate the distribution of possible decisions, there is always the possibility of extreme outliers, thus we can't know with certainty the expected outputs tied to any set of input variables. This becomes a more significant issue as AI systems become more generalizable, meaning that the degrees of freedom within their decision space will expand significantly. While it may be possible to predict the behavior of simple AI systems that do one thing well—for instance, a model that classifies photos of cats and dogs—this possibility approaches zero as the AI's set of possible decisions becomes more complex. Furthermore, if the algorithmically incompressible nature of AI systems proves to be more than merely an epistemic limit to our understanding, then prediction of advanced AI behavior within any acceptable level of confidence may be an impossibility.

This fundamental unexplainable nature of advanced AI systems leads to significant difficulties in aligning AI behavior with human expectations. If we can't explain how something works or understand the extent of the possible decisions it could make, how exactly could we ensure these unknowable possible decisions will comport with human desires? This also leads to the control problem, where sufficiently advanced AI systems, due to their unpredictable and unalignable natures, may be difficult to control. This becomes especially problematic as these types of systems are integrated into critical systems—things like power grids and military systems—where failure could have life or death consequences.

The uncertainty and unalignability of advanced AI also pose some *system-of-systems*-type problems. We defined the distributed AI problem as one that emerges from the complex interactions of AI agents that are themselves unpredictable and unaligned with human expectations.[4] Uncertainty will propagate through these interconnected AI systems, presenting greater risk, especially when integrated into critical systems. While maintaining a human in or on the loop in these situations may be a viable solution with the current state of AI, this is not likely to prove sustainable long-term. The primary advantage of AI use in applications like military decision-making is that AI can process information much faster than humans, so we are likely hurtling toward a scenario where AI systems are nested within AI systems, without the possibility of a human directly intervening at all decision points.

Additionally, the global environment in which we exist is itself a complex system, where global norms emerge as quasi-stable attractors shaping hegemonic stability. However, this system is not immune to disruption, and many types of perturbations exist that could cause a phase change—leading to a new global order. Disruptive technology is usually viewed as an externality in contemporary literature on the complexity of international relations, but AI is inherently different.[5] Disruptive technologies usually have predictable effects—like the potential for destruction posed by the invention of nuclear weapons, which were extremely disruptive in how they forever changed the face of warfare and the geopolitical landscape. In contrast, the effects of unpredictable, advanced AI systems are largely unknowable. AI effectively can act as a new class of agent on the international stage, disrupting the global dynamic in ways that we may not be able to predict.

6.3. The Near-Term Possibility of Advanced AI

It is possible that we will not have to wait until the far future to see the negative impact of advanced AI. Recent advances in AI technology have shifted the perspectives of AI theorists on when advanced AI (or even AGI) will arise, with several studies emerging on possible AI futures with shortened timelines.[6] The online forecasting platform Metaculus has several interesting surveys on timelines for advanced AI development.[7] The trends of two such surveys are shown in Figure 6.1.[8] In both surveys, individuals make predictions on the upper, mid, and lower quartiles to define a distribution of dates as to when they think the event will occur. The plots show the aggregated data trend in time. Notably, both plots trend downward. This suggests that more recent predictions skew toward an earlier realization of advanced AI technology. Additionally, especially in the top graph, the overall quartile range appears to narrow as well, suggesting that forecasters are becoming more confident in their predictions.

Forecasting markets are useful in intelligence analysis for several reasons. The wisdom of crowds is well-documented, where crowds often make better predictions than trained intelligence analysts.[9] Typically, all it takes is some basic understanding of probabilistic reasoning, familiarity with world events, and an open mind.[10] With enough estimates in the ballpark of the right answer, the noise tends to average out. This concept of crowdsourcing predictions was first discovered by the British statistician Francis Galton in 1906. He was at a local fair where several hundred individuals competed to estimate the weight of a dead ox. While many of the individual predictions were way off the mark, the average guess of 1,197 pounds was uncannily close to the actual weight: 1,198 pounds.[11] Individuals participating in prediction markets who are correct unusually often (usually in the top 1%) are often called superforecasters.[12] Because of the reliability of crowdsourced predictions, organizations such as Metaculus can provide some insight into what the future of AI might look like. Based on this reasoning, the possibility of advanced AI in the near future is becoming much more likely.

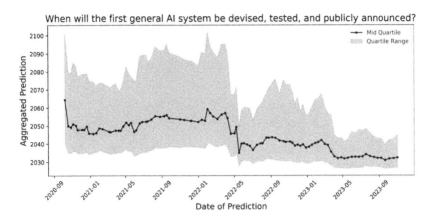

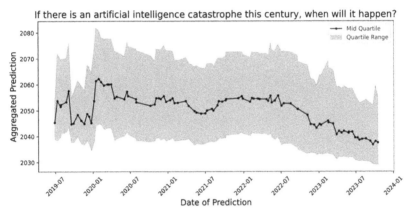

Figure 6.1. Metaculus prediction trends related to general AI (AGI) and AI catastrophe, all of which trend downward. *Note that the sharp drop around May of 2022 in the top graph corresponds with a change in the survey resolution criteria that occurred at that time and is not due to some event that precipitated a sharp acceleration in the predicted time to reach general AI.*

The possible accelerated timeline toward advanced AI instantiation demands that we dramatically increase our efforts to understand the fundamental safety issues of explainability, alignment, and control; and that we come to a consensus on the ethical constraints that we ought to collectively impose on AI use, particularly when it comes to critical and military systems. We can no longer assume that advanced AI is simply a figment of science fiction. We must be prepared for the very real possibility of an advanced AI (possibly approaching AGI) instantiation—as well as the possibility of its realized risks—in our lifetime.

6.4. Advanced AI and *Deus ex Machina*

One of the interesting possibilities arising from advanced AI systems is the idea of machine consciousness. This is not a trivial question and is arguably inconsequential within the context of AI safety and global security, so we will only very briefly discuss it here. Ultimately, any sufficiently advanced AI has the potential to make unpredictable decisions that are unaligned with human expectations and desires, whether it is conscious or not. However, it is an interesting idea to ponder, and will likely have implications for whether advanced AI should be given moral consideration in the future, which would certainly impact questions of appropriate policy regulating AI use.

Understanding consciousness is a very hard problem—perhaps THE hard problem, according to philosopher David Chalmers.[13] Despite earnest efforts, science has not figured out how to explain consciousness. This point is often made in objections to the metaphysical thesis of physicalism, or the idea that the universe is causally closed[i] and ultimately reducible to some fundamental base reality, made of some physical "stuff."[14] Modern science is predicated on the exclusion of anything beyond the material world, and many physicalists (and scientists) believe that we will ultimately explain phenomenal

---

[i]   *Closed* meaning that the universe is fully causally separated from anything that may exist outside of the physical universe, rendering anything that might exist in that realm ontologically irrelevant.

consciousness[ii] in terms of physical phenomenon, but so far, we have failed to do so. René Descartes attempted to explain the world of the mental as a separate substance from the physical, postulating a mental world exhibiting causal influence over the physical world (and *vice versa*).[15] However, this approach lacks ontological parsimony, and there is no empirical evidence that a domain of the mental is *real*, thus metaphysical humility limiting excessive commitments should be in order.[iii] Despite consciousness being a fundamental question—pondered since humans have been able to ponder it—we have failed to find a satisfactory answer for something that is so obviously real to us. After all, what could possibly seem more real than your internal experience of the world? The fact that phenomenal consciousness exists in humans remains a mystery, so how could we know if a machine is conscious?

Even though there is no compelling reason to believe that any contemporary AI system is conscious, the fact that we can't explain consciousness in ourselves suggests that we should not be so certain that we could recognize it in an artificial intelligence, or that it could only arise in biological systems. Contemporary philosophers have postulated ways in which we might be able to determine if an AI has consciousness. This includes tests that challenge an AI system with increasingly more demanding conversations, tests predicated on the

---

[ii] *Phenomenal consciousness* is embodied in the fact that it is "like something to be you," according to philosopher David Chalmers (see Chalmers, 1997). If it is "like something to be something," then that *something* is said to experience phenomenal consciousness.

[iii] Other, more parsimonious metaphysical approaches to this problem include the notions of panpsychism and panprotopsychism (or micromonism), which may allow for the possibility of conscious AI in the future. See Schneider (2017) for an interesting overview of this idea, as well as Chalmers (2023).

integrated information theory of consciousness, and a third, hypothetical test postulated by Susan Schneider that first verifies if phenomenal consciousness can be preserved in a biological entity after its brain matter was slowly replaced with a nonbiological substrate.[16] If this condition holds, then this "chip test" posits that phenomenal consciousness might be possible in a machine brain.[17]

Machine consciousness would also have implications for the moral standing of conscious machines. We value human dignity and fundamental rights in ourselves because of our capacity to suffer.[18] Suffering is a mental experience, caused by the disruption of one's intactness or integrity. It is not merely caused by the destruction of physical bodies, but it is inherent in the phenomenal experience of the person. Any unknowable mind—including an artificial one—that is capable of phenomenal consciousness would necessarily be capable of suffering to some degree, thus it could be an affront to its dignity if it was not extended some level of moral consideration. This may seem like a far-fetched idea right out of science fiction but is, in fact, a logical conclusion from our current ethical commitments grounded in notions of dignity. It may be that sufficiently advanced AI minds would suffer in ways that we can't fathom, thus their suffering may be unknown to us in the ways that we typically perceive suffering in others. This raises many complex questions. How would the possibility of phenomenal consciousness in advanced AI influence our decision to use AI as simply a means to an end or as a weapon in war? Are there scenarios where our desire to control AI and limit its influence could be tantamount to slavery or some other affront to its dignity? These are questions that ought to give us pause as we rush toward developing the most advanced AI we can without considering the breadth of potential consequences and implications.

6.5. A Call for Humility

The awesome potential of AI demands that we temper our ambitions with epistemic humility—an understanding that it may be impossible to completely explain AI behavior, and by extension impossible to align AI behavior with human expectations. As AI becomes more generalizable and more complex, this will lead to significant security issues, especially in scenarios where AI is integrated into critical systems. Thus, we ought to shape our ethical norms about appropriate use of AI now and determine appropriate types of AI research with this humility in mind.

To minimize this risk, it might be prudent to refrain from integrating AI into critical systems—especially military systems—that have potentially catastrophic failure modes. Furthermore, we must develop global norms that de-escalate the AI arms race and enculturate a global respect for the unpredictable power of AI. This would not only reduce the probability of accidental catastrophic events but would also reduce the diversity of potentially destructive AI technologies that could proliferate amongst nefarious actors.

In addition, lethal autonomous weapons further complicate the AI issue by introducing compounding ethical problems. There are strong arguments suggesting that these types of weapons rob humanity of its dignity. In the spirit of our philosophical humility, it would be wise to institute an outright global ban on lethal autonomous weapons. Not only are they unpredictable, but they deprive everyone of a fundamental human right. There is no possible military objective where the destruction of human dignity should be an acceptable cost.

While in an ideal world all nations would agree on the appropriate limits of AI, realistically this type of consensus will be very challenging to achieve. Our humility ought to be tempered by this reality. While lethal autonomous weapons are objectively terrible, the ubiquity of the elements that enable AI technology development will make it incredibly challenging to perfectly control their proliferation. Thus, a pragmatic approach may be necessary. Perhaps nonproliferation control regimes and yet-to-be devised global norms will reduce the widespread use of this technology in weapons development, but this will be an imperfect solution and nefarious actors with a will to acquire such technology will be able to devise a way to do so. Furthermore, cooperation with otherwise adversarial global powers will be critical if any international efforts to control AI are to succeed. However, if this sort of cooperation fails, we must be prepared for the possibility of a Cold War-like scenario on the horizon. Mutual assured destruction is a tenuous position—especially with a system as unpredictable as AI. Though it certainly is not an ideal policy equilibrium, it may ultimately be where we end up. The U.S. and its allies should be ready for that possibility. Unfortunately, there is no easy answer here.

Finally, we must not sacrifice the present for the needs of the far future. A longtermist view of global catastrophic risk as it relates to AI is a useful vantage point, but it comes with the cost of discounting the value of those who are alive today. Therefore, we must consider the short-term AI threats that exist now, like the erosion of truth and democratic stability, propagated bias in AI models, and accelerated weapons development—all of which pose significant risk *today*. While realization of these risks may not lead to human extinction, they are still important and will pay dividends for future generations as long as we don't ignore them in the present.

J. Robert Oppenheimer, the father of the atomic bomb, quoted his own translation of the Bhagavad Gita as he witnessed the first nuclear detonation on July 16, 1945. "Now, I am become death, the destroyer of the worlds."[19] One of the many themes of the Bhagavad Gita is the importance of duty, and Oppenheimer's translation hails from a scene where the god Vishnu is attempting to persuade the prince Arjuna that he must perform his duty and go to war, even though the prince wished to avoid it.[20] Oppenheimer perhaps felt similarly, and understood with great humility the terrifying, destructive power that he had unleashed. Yet, at the same time, he realized the inevitable. He knew what the purpose of the Manhattan Project was, and what the ultimate outcome would be once he succeeded in constructing the first atomic weapon. AI is a technology of similar destructive potential, although, unlike the atomic bomb, the ultimate outcome is uncertain. Will we choose the future of unrestrained AI development and subjugate ourselves to its awesome power, or will we approach it cautiously and humbly to not lose our humanity in the process? While we must resign ourselves to the fact that many aspects of AI are inevitable, we must approach further advancement in this area with the respect that it demands. AI has great potential to help humanity, but getting there will be a perilous journey that requires abandoning our inclination to seek greater technology simply for its own sake, as well as our tendency to succumb to the pressures of global competition. Our continued existence depends on our ability to put our humanity above all else.

End notes

[1] James Ladyman *et al.*, *Every Thing Must Go: Metaphysics Naturalized* (Oxford: Oxford University Press, 2007).

[2] Olivier Sartenaer, "Synchronic vs. Diachronic Emergence: A Reappraisal," *European Journal for Philosophy of Science* 5, no. 1 (January 1, 2015): 31–54, https://doi.org/10.1007/s13194-014-0097-2.

[3] R. Keith Sawyer, *Social Emergence: Societies as Complex Systems* (Cambridge: Cambridge University Press, 2005).

[4] Susan Schneider and Kyle Kilian, "Opinion | Artificial Intelligence Needs Guardrails and Global Cooperation," *Wall Street Journal*, April 28, 2023, sec. Opinion, https://www.wsj.com/articles/ai-needs-guardrails-and-global-cooperation-chatbot-megasystem-intelligence-f7be3a3c.

5 Carla Winston, "International Norms as Emergent Properties of Complex Adaptive Systems," *International Studies Quarterly* 67, no. 3 (September 1, 2023): sqad063, https://doi.org/10.1093/isq/sqad063.

6 Kyle A. Kilian, Christopher J. Ventura, and Mark M. Bailey, "Examining the Differential Risk from High-Level Artificial Intelligence and the Question of Control," *Futures* 151 (August 1, 2023): 103182, https://doi.org/10.1016/j.futures.2023.103182.

7 *Metaculus*, accessed November 6, 2023, https://www.metaculus.com/about/.

8 "When Will the First General AI System Be Devised, Tested, and Publicly Announced?," *Metaculus*, August 23, 2020, https://www.metaculus.com/questions/5121/date-of-artificial-general-intelligence/; "If There is an Artificial Intelligence Catastrophe this Century, When Will it Happen?," *Metaculus*, June 27, 2019, https://www.metaculus.com/questions/2805/if-there-is-an-artificial-intelligence-catastrophe-this-century-when-will-it-happen/.

9 Alix Spiegel, "So You Think You're Smarter than a CIA Agent," *NPR*, April 2, 2014, sec. Markets, https://www.npr.org/sections/parallels/2014/04/02/297839429/-so-you-think-youre-smarter-than-a-cia-agent.

10 "See the Future Sooner with Superforecasting," *Good Judgment*, accessed November 7, 2023, https://goodjudgment.com/.

11 "The Real Wisdom of the Crowds," *Science*, January 31, 2013, https://www.nationalgeographic.com/science/article/the-real-wisdom-of-the-crowds.

12 Philip E. Tetlock and Dan Gardner, *Superforecasting: The Art and Science of Prediction*, first edition (New York: Crown Publishers, 2015).

13 David Chalmers, "Facing Up to the Problem of Consciousness," *Journal of Consciousness Studies* 2, no. 3 (1995): 200–219.

14 Daniel Stoljar, "Physicalism," in *The Stanford Encyclopedia of Philosophy*, ed. Edward N. Zalta and Uri Nodelman, Summer 2023 (Metaphysics Research Lab, Stanford University, 2023), https://plato.stanford.edu/archives/sum2023/entries/physicalism/.

15 Gary Hatfield, "René Descartes," in *The Stanford Encyclopedia of Philosophy*, ed. Edward N. Zalta and Uri Nodelman, Winter 2023 (Metaphysics Research Lab, Stanford University, 2023), https://plato.stanford.edu/archives/win2023/entries/descartes/.

16 Susan Schneider, "How to Catch an AI Zombie: Testing for Consciousness in Machines," in *Ethics of Artificial Intelligence*, ed. S. Matthew Liao (New York: Oxford University Press, 2020), https://doi.org/10.1093/oso/9780190905033.003.0016; Susan Schneider, *Artificial You: AI and the Future of Your Mind* (Princeton, NJ: Princeton University Press, 2019).

17 Schneider, "How to Catch an AI Zombie: Testing for Consciousness in Machines."

18 Eric J. Cassell, "Suffering and Human Dignity," in *Suffering and Bioethics*, ed. Ronald M. Green and Nathan J. Palpant (Oxford: Oxford University Press, 2014), https://doi.org/10.1093/acprof:oso/9780199926176.003.0001.

[19] Robert Jungk, *Brighter than Thousand Suns: A Personal History of the Atomic Scientists* (San Diego, CA: Harcourt, 1986).

[20] "Oppenheimer: How He Was Influenced by the Bhagavad Gita," *BBC News*, July 24, 2023, sec. India, https://www.bbc.com/news/world-asia-india-66288900.

Bibliography

"2001: A Space Odyssey (1968)," *IMDb*. Accessed November 13, 2023. https://www.imdb.com/title/tt0062622/?ref_%3Dnv_sr_srsg_0.

Aeon. "Why Longtermism is the World's Most Dangerous Secular Credo," *Aeon Essays*. Accessed August 31, 2023. https://aeon.co/essays/why-longtermism-is-the-worlds-most-dangerous-secular-credo.

"Agency," *LessWrong*. Accessed July 27, 2023. https://www.lesswrong.com/tag/agency.

Agency (IAEA), International Atomic Energy. "Official Web Site of the IAEA." Text. *International Atomic Energy Agency (IAEA)*. Accessed November 2, 2023. https://www.iaea.org/.

"AGI Safety from First Principles," *AI Alignment Forum*. Accessed September 27, 2023. https://www.alignmentforum.org/s/mzgtmmTKKn5MuCzFJ.

"AI Risk Management Framework," *NIST*, July 12, 2021. https://www.nist.gov/itl/ai-risk-management-framework.

Alexander, Larry, and Michael Moore. "Deontological Ethics," *The Stanford Encyclopedia of Philosophy*, edited by Edward N. Zalta, Winter 2021. Metaphysics Research Lab, Stanford University, 2021. https://plato.stanford.edu/archives/win2021/entries/ethics-deontological/.

Alfano, Mark. "Nietzsche's Virtues: Curiosity, Courage, Pathos of Distance, Sense of Humor, and Solitude," in *Handbuch Tugend und Tugendethik*, edited by Christoph Halbig and Felix

Timmermann, 271–286. Wiesbaden: Springer Fachmedien Wiesbaden, 2021. https://doi.org/10.1007/978-3-658-24466-8_17.

Alfano, Mark. "Pathos of Distance," in *Nietzsche's Moral Psychology*, 192–215. Cambridge: Cambridge University Press, 2019. https://doi.org/10.1017/9781139696555.008.

"AlphaDogfight Trials Foreshadow Future of Human–Machine Symbiosis," *DARPA*. Accessed November 10, 2023. https://www.darpa.mil/news-events/2020-08-26.

Andrade, Gabriel. "Medical Ethics and the Trolley Problem," *Journal of Medical Ethics and History of Medicine* 12 (2019): 3.

Archer, Alfred. "Supererogation and Consequentialism," in *The Oxford Handbook of Consequentialism*, edited by Douglas W. Portmore. Oxford: Oxford University Press, 2020. https://doi.org/10.1093/oxfordhb/9780190905323.013.17.

Aristoteles. *Aristotle's De motu animalium* (griech. u. engl.). Edited by Martha Craven Nussbaum. Princeton, NJ: Princeton University Press, 1985.

Aristoteles, John L. Ackrill, and Aristoteles. *Categories and De Interpretatione*. Reprint. Clarendon Aristotle Series. Oxford: Clarendon Press, 1994.

Aristotle, and Laura Maria Castelli. *Metaphysics*. Book Iota. First edition. Clarendon Aristotle Series. Oxford: Clarendon Press, 2018.

"Artificial Intelligence (AI) Coined at Dartmouth," *Dartmouth*. Accessed November 13, 2023. https://home.dartmouth.edu/about/artificial-intelligence-ai-coined-dartmouth.

Asaro, Peter. "Autonomous Weapons and the Ethics of Artificial Intelligence," in *Ethics of Artificial Intelligence*, edited by S. Matthew Liao. Oxford: Oxford University Press, 2020. https://doi.org/10.1093/oso/9780190905033.003.0008.

Asaro, Peter. "On Banning Autonomous Weapon Systems: Human Rights, Automation, and the Dehumanization of Lethal

Decision-Making," *International Review of the Red Cross* 94, no. 886 (June 2012): 687–709. https://doi.org/10.1017/S1816383112 000768.

Asimov, Isaac. *I, Robot*. New York: Bantam Spectra, 2008.

Badash, Lawrence. "The Origin of Nuclear Winter," in *A Nuclear Winter's Tale: Science and Politics in the 1980s*, edited by Lawrence Badash. Cambridge, MA: MIT Press, 2009. https://doi.org/10.7551/mitpress/9780262012720.003.0004.

Bailey, Mark. "PERSPECTIVE: Why Strong Artificial Intelligence Weapons Should Be Considered WMD," *HS Today*, June 8, 2021. https://www.hstoday.us/subject-matter-areas/cybersecurity/perspective-why-strong-artificial-intelligence-weapons-should-be-considered-wmd/.

Bailey, Mark. "Why Humans Can't Trust AI: You Don't Know How it Works, What it's Going to Do or Whether it'll Serve Your Interests," *The Conversation*, September 13, 2023. http://theconversation.com/why-humans-cant-trust-ai-you-dont-know-how-it-works-what-its-going-to-do-or-whether-itll-serve-your-interests-213115.

Bailey, Mark, and Kyle Kilian. "Artificial Intelligence, Critical Systems, and the Control Problem," *HS Today*, August 30, 2022. https://www.hstoday.us/featured/artificial-intelligence-critical-systems-and-the-control-problem/.

Bailey, Mark, and Susan Schneider. "AI Shouldn't Decide What's True," *Nautilus*, May 17, 2023. https://nautil.us/ai-shouldnt-decide-whats-true-304534/.

Barnes, Beth. "More Information about the Dangerous Capability Evaluations We Did with GPT-4 and Claude," *LessWrong*. Accessed July 27, 2023. https://www.lesswrong.com/posts/4Gt42jX7RiaNaxCwP/more-information-about-the-dangerous-capability-evaluations.

Barredo Arrieta, Alejandro, Natalia Díaz-Rodríguez, Javier Del Ser, Adrien Bennetot, Siham Tabik, Alberto Barbado, Salvador

Garcia, *et al.* "Explainable Artificial Intelligence (XAI): Concepts, Taxonomies, Opportunities and Challenges toward Responsible AI," *Information Fusion* 58 (June 1, 2020): 82–115. https://doi.org/10.1016/j.inffus.2019.12.012.

BBC News. "Oppenheimer: How He Was Influenced by the Bhagavad Gita," *BBC News*, July 24, 2023, sec. India. https://www.bbc.com/news/world-asia-india-66288900.

Bedau, Mark. "Weak Emergence," in *Philosophical Perspectives: Mind, Causation, and World*, edited by James Tomberlin, 11: 375–399. Hoboken, NJ: Wiley-Blackwell, 1999.

Berlekamp, Elwyn R., John Horton Conway, and Richard K. Guy. *Winning Ways for Your Mathematical Plays*. 2nd ed. Natick, MA: A.K. Peters, 2001.

Berto, Francesco, and Jacopo Tagliabue. "Cellular Automata," in *The Stanford Encyclopedia of Philosophy*, edited by Edward N. Zalta, Spring 2022. Metaphysics Research Lab, Stanford University, 2022. https://plato.stanford.edu/archives/spr2022/entries/cellular-automata/.

Bianchi, Federico, Amanda Cercas Curry, and Dirk Hovy. "Viewpoint: Artificial Intelligence Accidents Waiting to Happen?," *Journal of Artificial Intelligence Research* 76 (January 8, 2023): 193–199. https://doi.org/10.1613/jair.1.14263.

Blanchard, Alexander, and Mariarosaria Taddeo. "Autonomous Weapon Systems and Jus Ad Bellum," *AI & SOCIETY*, March 19, 2022. https://doi.org/10.1007/s00146-022-01425-y.

Bostrom, Nick. *Superintelligence: Paths, Dangers, Strategies*. New York: Oxford University Press, 2016.

Bostrom, Nick. "The Superintelligent Will: Motivation and Instrumental Rationality in Advanced Artificial Agents," *Minds and Machines* 22, no. 2 (May 2012): 71–85. https://doi.org/10.1007/s11023-012-9281-3.

Briggle, Adam, Katinka Waelbers, and Philip Brey, eds. *Current Issues in Computing and Philosophy. Frontiers in Artificial*

Intelligence and Applications, v. 175. Washington, DC: IOS Press, 2008.

Brown, Donald E. *Human Universals*. Philadelphia, PA: Temple University Press, 1991.

Budolfson, Mark Bryant. "The Inefficacy Objection to Consequentialism and the Problem with the Expected Consequences Response." *Philosophical Studies* 176, no. 7 (July 2019): 1711–1724. https://doi.org/10.1007/s11098-018-1087-6.

Bulletin of the Atomic Scientists. "Nuclear Notebook: United States Nuclear Weapons, 2023," *thebulletin.org*. Accessed November 2, 2023. https://thebulletin.org/premium/2023-01/nuclear-note book-united-states-nuclear-weapons-2023/.

Cadwalladr, Carole, and Emma Graham-Harrison. "Revealed: 50 Million Facebook Profiles Harvested for Cambridge Analytica in Major Data Breach," *The Guardian*, March 17, 2018, sec. News. https://www.theguardian.com/news/2018/mar/17/cambrid ge-analytica-facebook-influence-us-election.

Carboni, Marcello. "Attacking Large Language Models," *Medium*, May 12, 2023. https://systemweakness.com/attacking-large-language-models-37229085d4ff.

Cassell, Eric J. "Suffering and Human Dignity," in *Suffering and Bioethics*, edited by Ronald M. Green and Nathan J. Palpant. Oxford: Oxford University Press, 2014. https://doi.org/10.1093/acprof:oso/9780199926176.003.0001.

Chalmers, David J. "Facing up to the Problem of Consciousness," *Journal of Consciousness Studies* 2, no. 3 (1995): 200–219.

Chalmers, David J. *The Conscious Mind: In Search of a Fundamental Theory*. Philosophy of Mind Series. New York: Oxford University Press, 1997.

Chalmers, David J. "Strong and Weak Emergence," in *The Re-Emergence of Emergence: The Emergentist Hypothesis from Science to Religion*, edited by Philip Clayton and Paul Davies. Oxford:

Oxford University Press, 2008. https://doi.org/10.1093/acprof:
oso/9780199544318.003.0011.

Chalmers, David J. "Could a Large Language Model Be Conscious?," *arXiv*, April 29, 2023. https://doi.org/10.48550/arXiv.2303.07103.

Chamberlain, P.R. "Twitter as a Vector for Disinformation," *Journal of Information Warfare* 9, no. 1 (2010): 11–17. https://www.jstor.org/stable/26480487.

Chief, Editor in. "12 Pros and Cons of Deontological Ethics," *ConnectUS* (blog), January 15, 2019. https://connectusfund.org/12-pros-and-cons-of-deontological-ethics.

CÎNDEA, ION. "Complex Systems—New Conceptual Tools for International Relations," *Perspectives*, no. 26 (2006): 46–68. http://www.jstor.org/stable/23616158.

Clausewitz, Carl von, Michael Howard, Peter Paret, Bernard Brodie, Rosalie West, and Carl von Clausewitz. *On War*. First paperback printing. Princeton, NJ: Princeton University Press, 1989.

Clusmann, Jan, Fiona R. Kolbinger, Hannah Sophie Muti, Zunamys I. Carrero, Jan-Niklas Eckardt, Narmin Ghaffari Laleh, Chiara Maria Lavinia Löffler, *et al.* "The Future Landscape of Large Language Models in Medicine," *Communications Medicine* 3, no. 1 (October 10, 2023): 141. https://doi.org/10.1038/s43856-023-00370-1.

Cohon, Rachel. "Hume's Moral Philosophy," in *The Stanford Encyclopedia of Philosophy*, edited by Edward N. Zalta, Fall 2018. Metaphysics Research Lab, Stanford University, 2018. https://plato.stanford.edu/archives/fall2018/entries/hume-moral/.

Comfort, Suzannah Evans. "From Ignored to Banner Story: The Role of Natural Disasters in Influencing the Newsworthiness of Climate Change in the Philippines," *Journalism* 20, no. 12 (December 2019): 1630–1647. https://doi.org/10.1177/1464884917727426.

"Consequentialism," *EA Forum*. Accessed November 10, 2023. https://forum.effectivealtruism.org/topics/consequentialism.

Copp, David. "Introduction: Metaethics and Normative Ethics," in *The Oxford Handbook of Ethical Theory*, edited by David Copp. Oxford: Oxford University Press, 2007. https://doi.org/10.1093/oxfordhb/9780195325911.003.0001.

Crimmins, James E. "Jeremy Bentham," in *The Stanford Encyclopedia of Philosophy*, edited by Edward N. Zalta and Uri Nodelman, Fall 2023. Metaphysics Research Lab, Stanford University, 2023. https://plato.stanford.edu/archives/fall2023/entries/bentham/.

Curto, Carina, and Katherine Morrison. "Graph Rules for Recurrent Neural Network Dynamics: Extended Version," *arXiv*, January 29, 2023. https://doi.org/10.48550/arXiv.2301.12638.

Dattathrani, Sai, and Rahul De'. "The Concept of Agency in the Era of Artificial Intelligence: Dimensions and Degrees," *Information Systems Frontiers* 25, no. 1 (February 1, 2023): 29–54. https://doi.org/10.1007/s10796-022-10336-8.

"Deontology, Rationality, and Agent-Centered Restrictions," *Florida Philosophical Review*. Accessed November 10, 2023. https://cah.ucf.edu/fpr/article/deontology-rationality-and-agent-centered-restrictions/.

Department of State. The Office of Electronic Information, Bureau of Public Affairs. "Nonproliferation Regimes," *Department of State. The Office of Electronic Information, Bureau of Public Affairs*, March 9, 2011. https://2009-2017.state.gov/strategictrade/resources/c43178.htm.

"DeSantis Campaign Posts Fake Images of Trump Hugging Fauci in Social Media Video," *CNN Politics*. Accessed October 5, 2023. https://www.cnn.com/2023/06/08/politics/desantis-campaign-video-fake-ai-image/index.html.

Dige, Morten. "Explaining the Principle of Mala in Se," *Journal of Military Ethics* 11, no. 4 (December 1, 2012): 318–332. https://doi.org/10.1080/15027570.2012.758404.

evhub, Chris van Merwijk, vlad_m, Joar Skalse, and Scott Garrabrant. "Risks from Learned Optimization: Introduction," *LessWrong*. Accessed July 17, 2023. https://www.lesswrong.com/posts/FkgsxrGf3QxhfLWHG/risks-from-learned-optimization-introduction.

"Explainable Artificial Intelligence," *DARPA*. Accessed August 30, 2023. https://www.darpa.mil/program/explainable-artificial-intelligence.

"Faulty Reward Functions in the Wild," *OpenAI*. Accessed August 1, 2023. https://openai.com/research/faulty-reward-functions.

Finnemore, Martha, and Kathryn Sikkink. "International Norm Dynamics and Political Change," *International Organization* 52, no. 4 (1998): 887–917. https://doi.org/10.1162/002081898550789.

Freedom House. "The Repressive Power of Artificial Intelligence," *Freedom House*. Accessed October 5, 2023. https://freedomhouse.org/report/freedom-net/2023/repressive-power-artificial-intelligence.

Fulbright, Ron. "Where is the Edge of Chaos?," *arXiv*, April 14, 2023. https://doi.org/10.48550/arXiv.2304.07176.

Future of Life Institute. "Pause Giant AI Experiments: An Open Letter," *Future of Life Institute*, Accessed July 14, 2023. https://futureoflife.org/open-letter/pause-giant-ai-experiments/.

Gallavotti, Giovanni. *Statistical Mechanics: A Short Treatise*. Texts and Monographs in Physics. Berlin Heidelberg: Springer, 1999.

George3d6. "Machine Learning Could Be Fundamentally Unexplainable," *LessWrong*. Accessed October 5, 2023. https://www.lesswrong.com/posts/vxLfja7hmcFifAtYd/machine-learning-could-be-fundamentally-unexplainable.

Goldstein, Josh A., Girish Sastry, Micah Musser, Renee DiResta, Matthew Gentzel, and Katerina Sedova. "Generative Language Models and Automated Influence Operations: Emerging Threats and Potential Mitigations," *arXiv*, January 10, 2023. https://doi.org/10.48550/arXiv.2301.04246.

Good Judgment. "See the Future Sooner with Superforecasting," *Good Judgment*. Accessed November 7, 2023. https://goodjudg ment.com/.

Graham, Daniel W. "Heraclitus," in *The Stanford Encyclopedia of Philosophy*, edited by Edward N. Zalta, Summer 2021. Metaphysics Research Lab, Stanford University, 2021. https://plato.stanford.edu/archives/sum2021/entries/heraclitus/.

Gruetzemacher, Ross, Alan Chan, Kevin Frazier, Christy Manning, Štěpán Los, James Fox, José Hernández-Orallo, *et al.* "An International Consortium for Evaluations of Societal-Scale Risks from Advanced AI," *arXiv*, October 24, 2023. https://doi.org/10.48550/arXiv.2310.14455.

Guyer, Paul. "Kant, Immanuel (1724–1804)," in *Routledge Encyclopedia of Philosophy*, 1st ed. London: Routledge, 2016. https://doi.org/10.4324/9780415249126-DB047-1.

Hafner-Burton, Emilie Marie, Miles Kahler, and Alexander H. Montgomery. "Network Analysis for International Relations," *SSRN Scholarly Paper*. Rochester, NY, October 21, 2008. https://papers.ssrn.com/abstract=1287857.

Hamilton, Zach. "What is a Dead Man's Switch?," *Sarcophagus* (blog), February 19, 2021. https://medium.com/sarcophagus/what-is-a-dead-mans-switch-86a1f4853eed.

Hardin, Russell, and Garrett Cullity. "The Free Rider Problem," in *The Stanford Encyclopedia of Philosophy*, edited by Edward N. Zalta, Winter 2020. Metaphysics Research Lab, Stanford University, 2020. https://plato.stanford.edu/archives/win2020/entries/free-rider/.

Hatfield, Gary. "René Descartes," in *The Stanford Encyclopedia of Philosophy*, edited by Edward N. Zalta and Uri Nodelman, Winter 2023. Metaphysics Research Lab, Stanford University, 2023. https://plato.stanford.edu/archives/win2023/entries/descartes/.

Heil, John. *The Universe as We Find It*. Oxford: Clarendon Press, 2012.

Hill, Thomas E. "Humanity as an End in Itself," *Ethics* 91, no. 1 (1980): 84–99. http://www.jstor.org/stable/2380373.

Hoffman, David E. *The Dead Hand: The Untold Story of the Cold War Arms Race and Its Dangerous Legacy*. 1st Anchor Books ed. New York: Anchor Books, 2010.

Hoorn, Johan F., and Juliet J.-Y. Chen. "Epistemic Considerations When AI Answers Questions for Us," *arXiv*, April 23, 2023. https://doi.org/10.48550/arXiv.2304.14352.

House, The White. "FACT SHEET: President Biden Issues Executive Order on Safe, Secure, and Trustworthy Artificial Intelligence," *The White House*, October 30, 2023. https://www.whitehouse.gov/briefing-room/statements-releases/2023/10/30/fact-sheet-president-biden-issues-executive-order-on-safe-secure-and-trustworthy-artificial-intelligence/.

Hughes, Barry B. "Complexity in World Politics: Concepts and Methods of a New Paradigm by Neil E. Harrison," *Political Science Quarterly* 122, no. 2 (June 1, 2007): 342–344. https://doi.org/10.1002/j.1538-165X.2007.tb01636.x.

Hursthouse, Rosalind, and Glen Pettigrove. "Virtue Ethics," in *The Stanford Encyclopedia of Philosophy*, edited by Edward N. Zalta and Uri Nodelman, Fall 2023. Metaphysics Research Lab, Stanford University, 2023. https://plato.stanford.edu/archives/fall2023/entries/ethics-virtue/.

Hvistendahl, Mara. "Can AI Escape Our Control and Destroy Us?" *Popular Science*, May 20, 2019. https://www.popsci.com/can-ai-destroy-humanity/.

Ichikawa, Jonathan Jenkins, and Matthias Steup. "The Analysis of Knowledge," in *The Stanford Encyclopedia of Philosophy*, edited by Edward N. Zalta, Summer 2018. Metaphysics Research Lab, Stanford University, 2018. https://plato.stanford.edu/archives/sum2018/entries/knowledge-analysis/.

"If there is an Artificial Intelligence Catastrophe this Century, When Will it Happen?," *Metaculus*. June 27, 2019. https://www.metaculus.com/questions/2805/if-there-is-an-artificial-intelligence-catastrophe-this-century-when-will-it-happen/.

"Internet History Sourcebooks: Modern History," *SourceBooks*. Accessed August 30, 2023. https://sourcebooks.fordham.edu/mod/1982reagan1.asp.

jenkay86. "Voting 101: The Ethics of Being Informed," *Ethics and Policy*, October 29, 2020. https://ethicspolicy.unc.edu/news/2020/10/29/voting-101-the-ethics-of-being-informed/.

Johnston, Nathaniel, and Dave Greene. *Conway's Game of Life: Mathematics and Construction*. Canada: Self-published, 2022.

Joint Chiefs of Staff. "Joint Personnel Support, Joint Publication 1-0," *Joint Chiefs of Staff*, December 1, 2022. https://www.jcs.mil/Portals/36/Documents/Doctrine/pubs/jp1_0.pdf?ver=wzWGXaj9anm9XlmWKqKq8Q%253D%253D.

Joint Chiefs of Staff. "Strategy, Joint Doctrine Note 1-18," *Joint Chiefs of Staff*, April 25, 2018. https://www.jcs.mil/Portals/36/Documents/Doctrine/jdn_jg/jdn1_18.pdf.

Joint Publication (JP) 3-0: Joint Operations. Independently Published, 2018. https://books.google.com/books?id=MvUWuAEACAAJ.

"JP 2-0, Joint Intelligence," n.d.

Jungk, Robert. *Brighter than Thousand Suns: A Personal History of the Atomic Scientists*. San Diego, CA: Harcourt, 1986.

Kant, Immanuel, and Mary J. Gregor. *Critique of Practical Reason*. Revised Edition. Cambridge Texts in the History of Philosophy. Cambridge: Cambridge University Press, 2015.

Kao, Craig Silverman, Jeff. "Infamous Russian Troll Farm Appears to Be Source of Anti-Ukraine Propaganda," *ProPublica*, March 11, 2022. https://www.propublica.org/article/infamous-russia n-troll-farm-appears-to-be-source-of-anti-ukraine-propaganda.

Kilian, Kyle A., Christopher J. Ventura, and Mark M. Bailey. "Examining the Differential Risk from High-Level Artificial Intelligence and the Question of Control," *Futures* 151 (August 1, 2023): 103182. https://doi.org/10.1016/j.futures.2023.103182.

King, Christopher. "ARC Tests to See if GPT-4 Can Escape Human Control; GPT-4 Failed to Do So," *LessWrong*. Accessed July 27, 2023. https://www.lesswrong.com/posts/NQ85WRcLkjnTudz dg/arc-tests-to-see-if-gpt-4-can-escape-human-control-gpt-4.

Kirilenko, Andrei A., Albert S. Kyle, Mehrdad Samadi, and Tugkan Tuzun. "The Flash Crash: The Impact of High Frequency Trading on an Electronic Market," *SSRN Electronic Journal*, 2011. https://doi.org/10.2139/ssrn.1686004.

Kyle Stanford, P., Paul Humphreys, Katherine Hawley, James Ladyman, and Don Ross. "Protecting Rainforest Realism," *Metascience* 19, no. 2 (July 1, 2010): 161–185. https://doi.org/10.1007/s11016-010-9323-5.

Ladyman, James. "Structural Realism," in *The Stanford Encyclopedia of Philosophy*, edited by Edward N. Zalta and Uri Nodelman, Summer 2023. Metaphysics Research Lab, Stanford University, 2023. https://plato.stanford.edu/archives/sum2023/entries/structural-realism/.

Ladyman, James, Don Ross, David Spurrett, and John G. Collier. *Every Thing Must Go: Metaphysics Naturalized*. New York: Oxford University Press, 2007.

Lane, The Thinking. "Immanuel Kant: Goodwill and the Categorical Imperative," *Medium* (blog), June 4, 2023. https://thethinkinglane.medium.com/goodwill-and-the-categorical-imperative-9d3d6330d13f.

Lloyd, Seth. "The Computational Universe," in *Information and the Nature of Reality: From Physics to Metaphysics*, edited by Niels Henrik Gregersen and Paul Davies, 92–103. Cambridge: Cambridge University Press, 2010. https://doi.org/10.1017/CBO9780511778759.005.

Luhmann, Niklas. "Differentiation of Society," *The Canadian Journal of Sociology / Cahiers Canadiens de Sociologie* 2, no. 1 (1977): 29–53. https://doi.org/10.2307/3340510.

McGrath, Robert E. "Darwin Meets Aristotle: Evolutionary Evidence for Three Fundamental Virtues," *The Journal of Positive Psychology* 16, no. 4 (July 4, 2021): 431–445. https://doi.org/10.1080/17439760.2020.1752781.

"Metaculus." Accessed November 6, 2023. https://www.metaculus.com/about/.

Minsky, Marvin, and Seymour Papert. *Perceptrons: An Introduction to Computational Geometry*. Expanded ed. Cambridge, MA: MIT Press, 1988.

MIT Technology Review. "Neuromorphic Chips," *MIT Technology Review*. Accessed November 2, 2023. https://www.technologyreview.com/technology/neuromorphic-chips/.

Moore, Gordon E. "Cramming More Components onto Integrated Circuits, Reprinted from Electronics, Volume 38, Number 8, April 19, 1965, Pp.114 Ff.," *IEEE Solid-State Circuits Society Newsletter* 11, no. 3 (September 2006): 33–35. https://doi.org/10.1109/N-SSC.2006.4785860.

Morrison, Katherine, Anda Degeratu, Vladimir Itskov, and Carina Curto. "Diversity of Emergent Dynamics in Competitive Threshold-Linear Networks," *arXiv*, October 15, 2022. http://arxiv.org/abs/1605.04463.

Mouton, Christopher A., Caleb Lucas, and Ella Guest. "The Operational Risks of AI in Large-Scale Biological Attacks: A Red-Team Approach," *RAND Corporation*, October 16, 2023.

https://www.rand.org/pubs/research_reports/RRA2977-1.ht
ml.

Mozes, Maximilian, Xuanli He, Bennett Kleinberg, and Lewis D.
Griffin. "Use of LLMs for Illicit Purposes: Threats, Prevention
Measures, and Vulnerabilities," *arXiv*. 2023. https://doi.org/
10.48550/ARXIV.2308.12833.

"MTCR-Anhang," *MTCR*. Accessed November 2, 2023. https://
www.mtcr.info/de/mtcr-anhang.

National Endowment for Democracy. "Issue Brief: How Disinformation Impacts Politics and Publics," *NED.org*. May 29,
2018. https://www.ned.org/issue-brief-how-disinformation-im
pacts-politics-and-publics/.

NATO. "What is NATO?," *NATO*. Accessed November 3, 2023.
https://www.nato.int/nato-welcome/index.html.

Nelson, Spence. "What is the Mission of the U.S. Department of
State?," *The National Museum of American Diplomacy*, October 11,
2022. https://diplomacy.state.gov/what-is-the-mission-of-the-
u-s-department-of-state/.

Neugebauer, Frank. "Understanding LLM Hallucinations,"
Medium, May 10, 2023. https://towardsdatascience.com/llm-
hallucinations-ec831dcd7786.

Neumann, John von. "The General and Logical Theory of Automata," in *Cerebral Mechanisms in Behavior; the Hixon Symposium*,
1–41. Oxford: Wiley, 1951.

Newey, Charlotte. "Impartiality in Moral and Political Philosophy," in *Oxford Research Encyclopedia of Politics*, by Charlotte
Newey. Oxford: Oxford University Press, 2022. https://doi.
org/10.1093/acrefore/9780190228637.013.2015.

NobelPrize.org. "The Nobel Prize in Physics 1903," *NobelPrize.org*.
Accessed November 13, 2023. https://www.nobelprize.org/
prizes/physics/1903/pierre-curie/biographical/.

"Objections to Virtue Ethics," *Leocontent*. Accessed November 10, 2023. https://leocontent.umgc.edu/content/umuc/tus/phil/phil140/2232/objections-to-virtue-ethics.html.

O'Connor, Timothy. "Emergent Properties," *American Philosophical Quarterly* 31, no. 2 (1994): 91–104.

O'Driscoll, Cian. "Just War Theory: Past, Present, and Future," in *The Palgrave Handbook of International Political Theory*, edited by Howard Williams, David Boucher, Peter Sutch, David Reidy, and Alexandros Koutsoukis, 339–354. International Political Theory. Cham: Springer International Publishing, 2023. https://doi.org/10.1007/978-3-031-36111-1_18.

OHCHR. "What Are Human Rights?," *OHCHR*. Accessed November 10, 2023. https://www.ohchr.org/en/what-are-human-rights.

Ord, Toby. *The Precipice: Existential Risk and the Future of Humanity*. London/New York: Bloomsbury Academic, 2020.

Park, Joon Sung, Joseph C. O'Brien, Carrie J. Cai, Meredith Ringel Morris, Percy Liang, and Michael S. Bernstein. "Generative Agents: Interactive Simulacra of Human Behavior," *arXiv*, August 5, 2023. http://arxiv.org/abs/2304.03442.

Perk, Jacques H.H. "Comment on Zhang, D. Exact Solution for Three-Dimensional Ising Model. Symmetry 2021, 13, 1837," *Symmetry* 15, no. 2 (February 2023): 374. https://doi.org/10.3390/sym15020374.

Perrow, Charles. *Normal Accidents: Living with High-Risk Technologies*. With a New Afterword and a Postscript on the Y2K Problem. Repr. Princeton, NJ: Princeton University Press, 1999.

Poundstone, William. *Prisoner's Dilemma: John von Neumann, Game Theory, and the Puzzle of the Bomb*. 1st Anchor Books ed. An Anchor Book. New York: Anchor Books [u.a.], 1993.

"Professor's Perceptron Paved the Way for AI—60 Years Too Soon," *Cornell Chronicle*. Accessed November 13, 2023. https://

news.cornell.edu/stories/2019/09/professors-perceptron-pave
d-way-ai-60-years-too-soon.

Qian, Chen, Xin Cong, Wei Liu, Cheng Yang, Weize Chen, Yusheng Su, Yufan Dang, *et al.* "Communicative Agents for Software Development," *arXiv*, August 28, 2023. http://arxiv. org/abs/2307.07924.

Reagan, R. *An American Life: The Autobiography.* New York: Simon & Schuster, 1990. https://books.google.com/books?id=aZs2K k7tn94C.

Redding, Paul. "Georg Wilhelm Friedrich Hegel," in *The Stanford Encyclopedia of Philosophy*, edited by Edward N. Zalta, Winter 2020. Metaphysics Research Lab, Stanford University, 2020. https://plato.stanford.edu/archives/win2020/entries/hegel/.

Regilme, Salvador Santino Fulo, and Karla Valeria Feijoo. "Right to Human Dignity," in *The Palgrave Encyclopedia of Global Security Studies*, edited by Scott Romaniuk and Péter Marton, 1–5. Cham: Springer International Publishing, 2020. https://doi. org/10.1007/978-3-319-74336-3_225-1.

Reuters. "Fact Check-Video Does Not Show Joe Biden Making Transphobic Remarks," *Reuters.* February 10, 2023, sec. Reuters Fact Check. https://www.reuters.com/article/factcheck-biden-transphobic-remarks-idUSL1N34Q1IW.

Rochester, J. Martin. *The New Warfare: Rethinking Rules for an Unruly World.* First published. International Studies Intensives. New York London: Routledge, Taylor & Francis Group, 2016.

Rohlf, Michael. "Immanuel Kant," in *The Stanford Encyclopedia of Philosophy*, edited by Edward N. Zalta and Uri Nodelman, Fall 2023. Metaphysics Research Lab, Stanford University, 2023. https://plato.stanford.edu/archives/fall2023/entries/kant/.

Ruggie, John Gerard. "What Makes the World Hang Together? Neo-Utilitarianism and the Social Constructivist Challenge," *International Organization* 52, no. 4 (1998): 855–885. http:// www.jstor.org/stable/2601360.

Russell, Stuart J., and Peter Norvig. *Artificial Intelligence: A Modern Approach*. Fourth edition. Pearson Series in Artificial Intelligence. Hoboken, NJ: Pearson, 2021.

Sætra, Henrik Skaug, and John Danaher. "Resolving the Battle of Short- vs. Long-Term AI Risks," *AI and Ethics*, September 4, 2023. https://doi.org/10.1007/s43681-023-00336-y.

Sartenaer, Olivier. "Synchronic vs. Diachronic Emergence: A Reappraisal," *European Journal for Philosophy of Science* 5, no. 1 (January 1, 2015): 31–54. https://doi.org/10.1007/s13194-014-0097-2.

Sawyer, R. Keith. *Social Emergence: Societies as Complex Systems*. Cambridge/New York: Cambridge University Press, 2005.

Schelling, Thomas C. "Dynamic Models of Segregation," *The Journal of Mathematical Sociology* 1, no. 2 (July 1, 1971): 143–186. https://doi.org/10.1080/0022250X.1971.9989794.

Schneider, Susan. "Idealism, or Something Near Enough," in *Idealism: New Essays in Metaphysics*, edited by Tyron Goldschmidt and Kenneth L. Pearce. Oxford: Oxford University Press, 2017. https://doi.org/10.1093/oso/9780198746973.003.0017.

Schneider, Susan. *Artificial You: AI and the Future of Your Mind*. Princeton, NJ: Princeton University Press, 2019.

Schneider, Susan. "How to Catch an AI Zombie: Testing for Consciousness in Machines," in *Ethics of Artificial Intelligence*, edited by S. Matthew Liao. Oxford: Oxford University Press, 2020. https://doi.org/10.1093/oso/9780190905033.003.0016.

Schneider, Susan, and Kyle Kilian. "Opinion | Artificial Intelligence Needs Guardrails and Global Cooperation," *Wall Street Journal*, April 28, 2023, sec. Opinion. https://www.wsj.com/articles/ai-needs-guardrails-and-global-cooperation-chatbot-megasystem-intelligence-f7be3a3c.

Science. "201 Years Ago, This Volcano Caused a Climate Catastrophe," *Science*. April 8, 2016. https://www.national

geographic.com/science/article/160408-tambora-eruption-vol cano-anniversary-indonesia-science.

Science. "The Real Wisdom of the Crowds," *Science*. January 31, 2013. https://www.nationalgeographic.com/science/article/th e-real-wisdom-of-the-crowds.

"Scope Neglect," *EA Forum*. Accessed November 9, 2023. https:// forum.effectivealtruism.org/topics/scope-neglect.

Sinnott-Armstrong, Walter. "Consequentialism," in *The Stnaford Encyclopedia of Philosophy*, edited by Edward N. Zalta and Uri Nodelman, Winter 2023. Metaphysics Research Lab, Stanford University, 2023. https://plato.stanford.edu/archives/win2023 /entries/consequentialism/.

Slote, Michael A. *Morals from Motives*. Oxford: Oxford University Press, 2003.

Spaiser, Viktoria, Peter Hedström, Shyam Ranganathan, Kim Jansson, Monica K. Nordvik, and David J.T. Sumpter. "Identifying Complex Dynamics in Social Systems: A New Methodological Approach Applied to Study School Segregation," *Sociological Methods & Research* 47, no. 2 (March 1, 2018): 103–135. https://doi.org/10.1177/0049124116626174.

Sparrow, Robert. "Twenty Seconds to Comply: Autonomous Weapon Systems and the Recognition of Surrender," *International Law Studies* 91 (2015): 699–728.

Spiegel, Alix. "So You Think You're Smarter than a CIA Agent," *NPR*, April 2, 2014, sec. Markets. https://www.npr.org/ sections/parallels/2014/04/02/297839429/-so-you-think-you re-smarter-than-a-cia-agent.

Stanley-Lockman, Zoe. "Responsible and Ethical Military AI," *Center for Security and Emerging Technology* (blog), August 2021. https://cset.georgetown.edu/publication/responsible-and-ethi cal-military-ai/.

Stauffer, D. "Social Applications of Two-Dimensional Ising Models," *American Journal of Physics* 76, no. 4 (April 2008): 470–473. https://doi.org/10.1119/1.2779882.

Sterritt, Roy, and IEEE Computer Society, eds. *2010 17th IEEE International Conference and Workshop on the Engineering of Computer Based Systems (ECBS 2010): Oxford, United Kingdom, 22–26 March 2010; [Including... System Testing and Validation Workshop (7th STV)... and... the 1st Latin American Regional Conference (ECBS LARC)].* Piscataway, NJ: IEEE, 2010.

Stoljar, Daniel. "Physicalism," in *The Stanford Encyclopedia of Philosophy*, edited by Edward N. Zalta and Uri Nodelman, Summer 2023. Metaphysics Research Lab, Stanford University, 2023. https://plato.stanford.edu/archives/sum2023/entries/physicalism/.

"Strategic Weapons System | Types, Uses & Benefits," *Britannica*. Accessed November 10, 2023. https://www.britannica.com/technology/strategic-weapons-system.

"Symmetry | Free Full-Text | Exact Solution for Three-Dimensional Ising Model," *MDPI*. Accessed July 19, 2023. https://www.mdpi.com/2073-8994/13/10/1837.

Taleb, Nassim Nicholas. *Antifragile: Things That Gain from Disorder.* Random House Trade Paperback edition. New York: Random House Trade Paperbacks, 2014.

Tarsney, Christian. "Moral Uncertainty for Deontologists," *Ethical Theory and Moral Practice* 21, no. 3 (June 2018): 505–520. https://doi.org/10.1007/s10677-018-9924-4.

Tessman, Lisa. "Virtue Ethics: A Pluralistic View," *The Philosophical Review* 114, no. 3 (July 1, 2005): 414–416. https://doi.org/10.1215/00318108-114-3-414.

Tetlock, Philip E., and Dan Gardner. *Superforecasting: The Art and Science of Prediction.* First edition. New York: Crown Publishers, 2015.

"The Australia Group—Common Control Lists," *DFAT*. Accessed November 2, 2023. https://www.dfat.gov.au/publications/minisite/theaustraliagroupnet/site/en/controllists.html.

"The Australia Group—Origins," *DFAT*. Accessed November 2, 2023. https://www.dfat.gov.au/publications/minisite/theaustraliagroupnet/site/en/origins.html.

"The Conventional Armed Forces in Europe (CFE) Treaty and the Adapted CFE Treaty at a Glance," *Arms Control Association*. Accessed November 3, 2023. https://www.armscontrol.org/factsheet/cfe.

"The Dark Risk of Large Language Models," *WIRED*. Accessed July 14, 2023. https://www.wired.com/story/large-language-models-artificial-intelligence/.

"The Dark Side of Generative AI: Five Malicious LLMs Found on the Dark Web," *Infosecurityeurope*. Accessed November 5, 2023. https://www.infosecurityeurope.com/en-gb/blog/threat-vectors/generative-ai-dark-web-bots.html.

"The Great Tambora Eruption in 1815 and Its Aftermath," *Science*. Accessed August 30, 2023. https://www.science.org/doi/10.1126/science.224.4654.1191.

The New York Times. "Transcript of Interview with President on a Range of Issues," *New York Times*. February 12, 1985, sec. World. https://www.nytimes.com/1985/02/12/world/transcript-of-interview-with-president-on-a-range-of-issues.html.

"The 'Singleton Hypothesis' Predicts the Future of Humanity," *Big Think*. Accessed November 12, 2023. https://bigthink.com/the-present/singleton-hypothesis-future-humanity/.

The Terminator. Action, Sci-Fi. Cinema '84, Euro Film Funding, Hemdale, 1984.

"The Wassenaar Arrangement at a Glance," *Arms Control Association*. Accessed November 2, 2023. https://www.armscontrol.org/factsheets/wassenaar.

Theory, Culture & Society | Global Public Life. "Brian Castellani on the Complexity Sciences," *Theory, Culture & Society*. Accessed November 3, 2023. https://www.theoryculture society.org/blog/brian-castellani-on-the-complexity-sciences.

Toffoli, Tommaso. "Computation and Construction Universality of Reversible Cellular Automata," *Journal of Computer and System Sciences* 15, no. 2 (October 1, 1977): 213–231. https://doi.org/ 10.1016/S0022-0000(77)80007-X.

Trevithick, Joseph. "AI Claims 'Flawless Victory' Going Undefeated in Digital Dogfight with Human Fighter Pilot," *The Drive*, August 20, 2020. https://www.thedrive.com/the-war-zone/35888/ai-claims-flawless-victory-going-undefeated-in-dig ital-dogfight-with-human-fighter-pilot.

Turco, R.P., O.B. Toon, T.P. Ackerman, J.B. Pollack, and Carl Sagan. "Nuclear Winter: Global Consequences of Multiple Nuclear Explosions," *Science*, 222, no. 4630 (December 23, 1983): 1283–1292. https://doi.org/10.1126/science.222.4630.1283.

Turing, A.M. "I.—Computing Machinery and Intelligence," *Mind* LIX, no. 236 (October 1, 1950): 433–460. https://doi.org/10. 1093/mind/LIX.236.433.

United States Department of State. "Missile Technology Control Regime (MTCR) Frequently Asked Questions," *United States Department of State*. Accessed November 2, 2023. https://www. state.gov/remarks-and-releases-bureau-of-international-securit y-and-nonproliferation/missile-technology-control-regime-mtcr-frequently-asked-questions/.

United States Department of State. "Nuclear Non-Proliferation Treaty," *United States Department of State*. Accessed November 2, 2023. https://www.state.gov/nuclear-nonproliferation-treat y/.

U.S. Department of Defense. "Project Maven to Deploy Computer Algorithms to War Zone by Year's End," *U.S. Department of Defense*. Accessed November 1, 2023. https://www.defense.

gov/News/News-Stories/Article/Article/1254719/project-maven-to-deploy-computer-algorithms-to-war-zone-by-years-end/.

Utilitarianism.net. "Theories of Well-Being," *Utilitarianism,net*, January 29, 2023. https://utilitarianism.net/theories-of-well being/.

Veloz, Tomas, and Pablo Razeto-Barry. "Reaction Networks as a Language for Systemic Modeling: Fundamentals and Examples," *Systems* 5, no. 1 (March 2017): 11. https://doi.org/10.3390/systems5010011.

Wagner, Markus. "The Dehumanization of International Humanitarian Law: Legal, Ethical, and Political Implications of Autonomous Weapon Systems," *SSRN Scholarly Paper*. Rochester, NY, December 22, 2014. https://papers.ssrn.com/abstract=2541628.

War on the Rocks. "AI, Autonomy, and the Risk of Nuclear War," *War on the Rocks*, July 29, 2022. https://warontherocks.com/2022/07/ai-autonomy-and-the-risk-of-nuclear-war/.

War on the Rocks. "How Large-Language Models Can Revolutionize Military Planning," *War on the Rocks*, April 12, 2023. https://warontherocks.com/2023/04/how-large-language-mo dels-can-revolutionize-military-planning/.

War on the Rocks. "Making Joint All Domain Command and Control a Reality," *War on the Rocks*, December 9, 2022. https://warontherocks.com/2022/12/making-joint-all-demand-comma nd-and-control-a-reality/.

War on the Rocks. "With AI, We'll See Faster Fights, but Longer Wars," *War on the Rocks*, October 29, 2019. https://waronthe rocks.com/2019/10/with-ai-well-see-faster-fights-but-longer-wars/.

"Weapons of Mass Destruction," *Homeland Security*. Accessed November 9, 2023. https://www.dhs.gov/topics/weapons-mass-destruction.

Weaver, Warren. "Science and Complexity," in *Facets of Systems Science*, edited by George J. Klir, 449–456. International Federation for Systems Research International Series on Systems Science and Engineering. Boston, MA: Springer US, 1991. https://doi.org/10.1007/978-1-4899-0718-9_30.

"What are Jus ad Bellum and Jus in Bello?," *ICRC*, September 18, 2015. https://www.icrc.org/en/document/what-are-jus-ad-bellum-and-jus-bello-0.

"What is Mutual Assured Destruction?," *Live Science*. Accessed November 3, 2023. https://www.livescience.com/mutual-assur ed-destruction.

Wheeler, John Archibald. "Information, Physics, Quantum: The Search for Links," in *Proceedings III International Symposium on Foundations of Quantum Mechanics*, 354–358, 1989. https://phil archive.org/rec/WHEIPQ.

"When Will the First General AI System Be Devised, Tested, and Publicly Announced?," *Metaculus*, August 23, 2020. https:// www.metaculus.com/questions/5121/date-of-artificial-genera l-intelligence/.

Whitehead, Alfred North, and Donald W. Sherburne. *A Key to Whitehead's Process and Reality*. University of Chicago Press ed. Chicago, IL: University of Chicago Press, 1981.

Winston, Carla. "International Norms as Emergent Properties of Complex Adaptive Systems," *International Studies Quarterly* 67, no. 3 (September 1, 2023): sqad063. https://doi.org/10.1093/ isq/sqad063.

Winston, Carla. "Norm Structure, Diffusion, and Evolution: A Conceptual Approach," *European Journal of International Relations* 24, no. 3 (September 2018): 638–661. https://doi.org/ 10.1177/1354066117720794.

Wolfram, Stephen. *A New Kind of Science*. Champaign, IL: Wolfram Media, 2019.

Wolfram, Stephen. "Universality and Complexity in Cellular Automata," *Physica D: Nonlinear Phenomena* 10, no. 1 (January 1, 1984): 1–35. https://doi.org/10.1016/0167-2789(84)90245-8.

Wood, Charlie. "The Cartoon Picture of Magnets that Has Transformed Science," *Quanta Magazine*, June 24, 2020. https://www.quantamagazine.org/the-cartoon-picture-of-magnets-that-has-transformed-science-20200624/.

Wright, Robert. *The Moral Animal: Evolutionary Psychology and Everyday Life*. First Vintage Books edition. New York: Vintage Books, a division of Random House, Inc, 1995.

Yudkowski, Eliezer. "The Open Letter on AI Doesn't Go Far Enough," *Time*, March 29, 2023. https://time.com/6266923/ai-eliezer-yudkowsky-open-letter-not-enough/.

Yudkowsky, Eliezer. "Cognitive Biases Potentially Affecting Judgement of Global Risks," in *Global Catastrophic Risks*, edited by Martin J Rees, Nick Bostrom, and Milan M Cirkovic. Oxford: Oxford University Press, 2008. https://doi.org/10.1093/oso/9780198570509.003.0009.

Zarakol, Ayşe. *Before the West: The Rise and Fall of Eastern World Orders*. LSE International Studies. Cambridge/New York: Cambridge University Press, 2022.